DISPOSED OF
BY LIBRARY
HOUSE OF LORDS

UNEP Studies   Volume 1

The Effects of Weapons on Ecosystems

UNEP Studies

Volume 1	The Effects of Weapons on Ecosystems J. P. Robinson
Volume 2	Environment and Development in Africa Environment Training Programme (ENDA)
Forthcoming	The Child in the African Environment R. O. Ohuche and B. Otaala

A previous publication within the scope of this series was published directly by UNEP, but is available from Pergamon Press:

TECHNOLOGY, DEVELOPMENT AND THE ENVIRONMENT: A RE-APPRAISAL
A. K. N. Reddy

NOTICE TO READERS

Dear Reader

If your library is not already a standing order customer or subscriber to this UNEP series may we recommend that you place a standing or subscription order to receive immediately upon publication all new volumes published in this valuable series. Should you find that these volumes no longer serve your needs your order can be cancelled at any time without notice.

ROBERT MAXWELL
Publisher at Pergamon Press

The Effects of Weapons on Ecosystems

J. P. ROBINSON
Science Policy Research Unit, University of Sussex, Brighton, UK

WITH THE ASSISTANCE OF
the United Nations Environment Programme
and
the United Nations Centre for Disarmament

Published for the
United Nations Environment Programme
by
PERGAMON PRESS
OXFORD · NEW YORK · TORONTO · SYDNEY · PARIS · FRANKFURT

U.K.	Pergamon Press Ltd., Headington Hill Hall, Oxford OX3 0BW, England
U.S.A.	Pergamon Press Inc., Maxwell House, Fairview Park, Elmsford, New York 10523, U.S.A.
CANADA	Pergamon of Canada, Suite 104, 150 Consumers Road, Willowdale, Ontario M2J 1P9, Canada
AUSTRALIA	Pergamon Press (Aust.) Pty. Ltd., P.O. Box 544, Potts Point, N.S.W. 2011, Australia
FRANCE	Pergamon Press SARL, 24 rue des Ecoles, 75240 Paris, Cedex 05, France
FEDERAL REPUBLIC OF GERMANY	Pergamon Press GmbH, 6242 Kronberg-Taunus, Pferdstrasse 1, Federal Republic of Germany

Copyright © 1979 UNEP and UN Centre for Disarmament

All Rights Reserved. No part of this publication may be reproduced, stored in a retrieval system or transmitted in any form or by any means: electronic, electrostatic, magnetic tape, mechanical, photocopying, recording or otherwise, without permission in writing from the copyright holders.

First edition 1979

British Library Cataloguing in Publication Data
Robinson, Julian Perry
The effects of weapons on ecosystems - (United Nations Environment Programme. UNEP studies).
1. Arms and armor - Environmental aspects
I. Title II. Series
574.5'222   QH545.A/   79-41226
ISBN 0 08 025656 2

The designations employed and the presentation of material in this publication do not imply the expression of any opinion whatsoever on the part of the secretariat of the United Nations Environment Programme concerning the legal status of any country, territory, city or area or of its authorities, or concerning the delineation of its frontier or boundaries.

Printed in Great Britain by A. Wheaton & Co., Ltd., Exeter

CONTENTS

INTRODUCTION vii

CHAPTER 1. BACKGROUND AND APPROACH
    Background 1
    Approach 6

CHAPTER 2. WEAPONS
    Classification of weapons 11
    High-explosive weapons 16
    Incendiary weapons 17
    Chemical weapons 18
    Biological weapons 22
    Nuclear weapons 24

CHAPTER 3. ECOSYSTEMS
    Structural and functional components 28
    Adaptability and vulnerability 31
    Classification of ecosystems 33
    Arid ecosystems 35
    Tropical ecosystems 37
    Arctic ecosystems 39
    Insular ecosystems 40
    Temperate ecosystems 40

CHAPTER 4. POSSIBLE IMPACTS OF WEAPONS ON ECOSYSTEMS

Empirical data	42
Theoretical framework	46
Stress through soil damage	47
Stress through plant-cover destruction	49
Stress through biocide	52
Relationships between scales of weapons-use and magnitudes of stress and strain	53
Impacts on tropical ecosystems	56
Impacts on arid ecosystems	58
Impacts on arctic ecosystems	60

CHAPTER 5. CONCLUSIONS AND RECOMMENDATIONS

Overall implications	62
Desertification	63
Magnitude of impact	65
Further study	66
Recommendations	68

TABLES

1. Classification of weapons	12
2. Effects of ground-burst nuclear weapons	25
on different components of ecosystems	25
3. Subsystems of the global ecosystem	34
4. Ecosystem-stressing potentials of different weapons	55

ANNEX

Field investigations into the ecological sequelae of military herbicide operations in Indochina: A bibliography	70

INTRODUCTION

This study brings together two seemingly separate but interrelated topics of growing concern to the international community: the destruction by man of his environment and its various ecosystems; and the harmful effects—both intended and unanticipated—of the production, testing, stockpiling and use of weapons of mass destruction.

These two concerns recently found expression at the UN Conference on Desertification (UNCOD), which was held at Nairobi from 29 August to 9 September 1977, attended by some 500 delegates from 94 countries and by various organizations. After extensive discussion at its plenary session and Committee of the Whole, the Conference adopted as resolution 4, "Effect of weapons of mass destruction on ecosystems". The resolution condemned the use of any techniques that caused the destruction of the environment, denounced the effects of destructive weapons and practices on ecosystems and condemned the use of poisons and of chemical and biological weapons. It also appealed to all States members of the organizations of the UN system to refrain from using or supplying to those who support policies of destruction, arms or chemical products for military use that have a widespread, long-lasting or severe effect on the environment. Finally, the Conference requested the Secretary-General to report on the implementation of the resolution to the UN General Assembly.

The thirty-second session of the General Assembly, which took place shortly after UNCOD, asked the Secretary-General of the United Nations to report on the implementation of the resolutions adopted by the Conference and to submit a report on this subject to its thirty-third session, with emphasis on resolution 4 and another resolution on financial and technical assistance to the least developed countries, resolution 2. In accordance with the General Assembly's request and taking account of the discussions held and appeals made both at the Conference and at the General Assembly, the Secretary-General of the United Nations commissioned a study of the effect of weapons of mass destruction on ecosystems. Because this was a subject of common interest, the UN Enviroment Programme and the UN Centre for Disarmament held consultations on how best to respond to the General Assembly's request.

To help prepare the Secretary-General's report to the General Assembly, Mr. Julian Perry Robinson, Senior Fellow of the Science Policy Research Unit at the University of Sussex, was retained as a Consultant. In co-operation with the UN Centre for Disarmament and the staff of UNEP, especially its New York Liaison Office, he undertook a study of the subject, consulting with other experts on various aspects of weapons and ecosystems and examining available literature. A short, condensed report was presented to the thirty-third session of the General Assembly as a special annex to the Secretary-General's report on the implementation of the resolutions adopted by UNCOD (A/33/259, Annex II). In view of the interest expressed in this subject, it was decided that a fuller report should be published. Arrangements were made for a further review of the texts of the study by the Centre for Disarmament, by UNEP and by a panel of eminent experts on the different aspects of ecology. They included the following:

Dr. René J. Dubos
Professor Emeritus
Rockefeller University

Professor Arthur Galston
Department of Botany
Yale University

Dr. M. El-Kassas
Professor of Ecology
Faculty of Science
University of Cairo

Professor Sanga Sabhasri
Secretary-General
National Rearch Council of Thailand

INTRODUCTION

Dr. M. S. Swaminathan
Director-General
Indian Council of Agricultural Research

Dr. Gilbert White
Former Director
Institute of Behavioural Science University of Colorado

The report that follows was prepared by Mr. Robinson, assisted by staff of UNEP and the Centre for Disarmament, after the comments of the panel of consultants. It is hoped that this study will respond to the concern of the international community over the effects of weapons of war and the destruction of ecosystems. As the study shows, there is still a large body of missing knowledge and much work remains to be done to be able to answer the critical questions raised. It is hoped, however, that the present study will clarify some of the outstanding issues to concerned individuals, stimulate follow-up studies to advance the cause of disarmament and the protection of the environment, and lead to further action on these two very important and interrelated subjects.

ROLF G. BJÖRNERSTEDT
Assistant Secretary-General
UN Centre for Disarmament

MOSTAFA K. TOLBA
Executive Director
UN Environment Programme

Chapter 1

BACKGROUND AND APPROACH

BACKGROUND

Particularly important in the preparatory work for the 1977 UN Conference on Desertification (UNCOD)[1] were the investigations that had been made into the causes of desertification. Such research had been in progress for many years, but recent events had greatly increased its urgency, stimulating also the call for international co-operation in combating desertification to which the Conference itself was a response. By 1973 the Sahelian region of Africa, lying in the southern margin of the Sahara, had been afflicted by five continuous years of drought. Two-thirds of Lake Chad had disappeared, and neither the Niger nor the Senegal Rivers had flooded, which meant that much of the best agricultural land of the region had become barren. Shallow and seasonal wells had dried up. Vegetation had been lost to hungry animals stripping the land. New patches of desert seemed to grow and link with the great desert to the north.[2] Thousands upon thousands of people had been forced to flee the region, losing their traditional means of livelihood and, a great many of them, their lives.

Was such a catastrophe to be ascribed solely to natural causes, against which human endeavour was impotent? Or had human and societal factors contributed to such an extent that their regulation might ameliorate, even prevent, future catastrophe? Such questions of causation are extremely complex,

---

[1] Some of the preparatory background documents have been published by Pergamon Press in association with UNEP as a consolidated book, *Desertification: Its Causes and Consequences*; for others, similar arrangements are being made by UNEP and UNESCO.
[2] *UN Conference on Desertification: Round-up, Plan of Action and Resolutions* (New York: United Nations, Office of Public Information, CESI.E52, March 1978), p. 1.

requiring investigation at several levels of analysis, using the tools of natural, social and political scientists. The knowledge acquired thus far has increased understanding in some areas of the subject. No less important, it has also increased awareness of ignorance in other areas. There is, in particular, the question of whether the mechanisms of desertification at work in the Sahel are representative of those that might operate in other places or at other times. Can different forms of desertification be envisaged that might prove no less disastrous in their impact on human life and socio-economic development? And, going further, may there not be other forms of interaction between human societies and their natural environment which could have consequences no less grave than those of desertification? Such considerations arose automatically at the time of UNCOD from the growing concerns that had found expression in the Declaration of the 1972 UN Conference on the Human Environment.

Broader issues of this type duly received attention in the course of UNCOD and in the preparatory work for it. Among them was the matter of war and the burgeoning potential for mass destruction represented by modern weapons technology. In the course of one of the UNCOD preparatory meetings, the Regional Preparatory Meeting for Africa South of the Sahara (Nairobi, 12 to 16 April 1977), held jointly with the Symposium on Drought and Desertification in Africa convened by the Organization of African Unity, particular reference was made to the role of chemical and biological weapons in desertification, an issue which was felt to be actively present in Africa.[3] These weapons, especially chemical means of deforestation, were also discussed in one of the background documents for the Conference, *Desertification: An Overview*.[4] The Conference ultimately voiced these concerns in the fourth of the eight resolutions which it adopted, the operative paragraphs[5] being as follows:

"1. *Condemns* the use of any techniques that cause the destruction of the environment;

"2. *Denounces* the effects of destructive weapons and practices on the ecosystems of all countries which have suffered them, and particularly developing countries, including those which are currently

---

[3] UNCOD. Reports of the Regional Preparatory Meetings, A/CONF.74/33/Add.1, at para. 107.
[4] A/CONF.74/1/Rev.1, at para. 197.
[5] The full text of UNCOD resolution 4, adopted on 9 September 1977, is to be found in the *Round-up* (*supra*, Note 2), at p. 40.

BACKGROUND AND APPROACH

engaged in the struggle for independence and those which have recently achieved independence through armed struggle;

"3. *Condemns* further the use of chemical and biological weapons which destroy or diminish the potential of ecosystems and are conducive to desertification;

"4. *Condemns*, and demands the prohibition of, the use of poisons in water as a weapon of war;

"5. *Appeals* to all States members of the organizations of the United Nations system to refrain from using or supplying to those who support this policy of destruction, arms or chemical products for military use that have a widespread, long-lasting or severe effect on the environment;

"6. *Requests* the Secretary-General to report on the implementation of the present resolution to the General Assembly."

Some 14 weeks later, the UN General Assembly adopted a resolution (UN Conference on Desertification, 32/172 of 19 December 1977) which, among other things, requested the Secretary-General to report to the Assembly at its next session on the implementation of the UNCOD resolutions.

Operative paragraph 5 of UNCOD resolution 4 implied that there existed a body of knowledge sufficient to differentiate "arms or chemical products for military use that have a widespread, long-lasting or severe effect on the environment" from those that do not have such effects. A study of this and of related matters raised in the resolution was commissioned by the Executive Director of the UN Environment Programme in consultation and co-operation with the UN Centre for Disarmament in order to assist the Secretary-General in responding to General Assembly resolution 32/172. A brief summary of this study was included as an annex to the report which the Secretary-General duly made to the General Assembly at its thirty-third session.[6] The present publication is the full version of the study, modified to reflect comments made upon it by authorities in the field who were subsequently consulted by UNEP. The broad findings and conclusions remain unchanged.

UNCOD resolution 4 may be seen as a component of the endeavours that have been increasing steadily within the international community to strengthen and extend measures for protecting the environment during time of war. Principle 26 of the UN Declaration on the Human Enviroment of 1972[7]

[6] A/33/259 of 20 October 1978.
[7] UN Environment Programme. *Compendium of Legislative Authority* at p. 11.

provided an important stimulus, as did the 1972 UNESCO Convention concerning the Protection of World Cultural and Natural Heritage.[8] Further concrete measures were taken in 1977. The first was the opening for signature of the Convention on the Prohibition of Military or any other Hostile Use of Environmental Modification Techniques,[9] a convention which entered into force in October 1978; the uses of environmental modification techniques which it prohibits are those having "widespread, long lasting or severe effects", and are understood to include the use of chemical herbicides having such effects. The second concrete measure was the opening for signature in December 1977, following negotiations during the Geneva Diplomatic Conference on the Reaffirmation and Development of International Humanitarian Law Applicable in Armed Conflict, of two Additional Protocols to the 1949 Geneva Conventions; Article 35 (3) of Additional Protocol I prohibits "methods or means of warfare which are intended, or may be expected, to cause widespread, long-term and severe damage to the natural environment", and Article 55 (2) prohibits attacks against the natural enviroment by way of reprisals.[10]

There are several other international agreements which, by virtue of restrictions which they place on possession, use or deployment of certain categories of weapon, are also relevant. Thus, the 1959 Antarctic Treaty[11] prohibits any measures of a military nature within Antarctica; the 1967 Treaty on Principles governing the Activities of States in the Exploration and use of Outer Space, including the Moon and other Celestial Bodies[12] outlaws the stationing of any kind of weapon of mass destruction in orbit around the earth or anywhere else in outer space; the 1967 Treaty for the Prohibition of Nuclear Weapons in Latin America[13] establishes a nuclear-weapons-free-zone within that region; and there is the 1971 Treaty on the Prohibition of the Emplacement of Nuclear Weapons and other weapons of Mass Destruction on the Seabed and the Ocean Floor and in the Subsoil Thereof.[14] Of special

---

[8] *UN Juridical Yearbook 1972* (New York: United Nations, Sales No. E.74.V.1, 1974), at pp. 89–99.
[9] United Nations. *Status of Multilateral Arms Regulation and Disarmament Agreements.* (New York: United Nations, Sales No. E.78.IX.2, 1978), at pp. 130–134.
[10] A/32/144 (15 August 1977).
[11] United Nations, *Treaty Series*, Vol. 402 (1962).
[12] *Ibid.*, Vol. 610 (1970).
[13] *Ibid.*, Vol. 634 (1970).
[14] *Supra*, note 9, at pp. 97–101.

relevance is the 1972 Convention on the Prohibition of the Development, Production and Stockpiling of Bacteriological (Biological) and Toxin Weapons, and on their Destruction.[15] The use of such weapons, and of chemical weapons, is proscribed by the Protocol for the Prohibition of the Use in War of Asphyxiating, Poisonous or Other Gases, and of Bacteriological Methods of Warfare, signed at Geneva on 17 June 1925.[16] It is to be noted, however, that varying understandings exist as to the precise scope of the prohibition of chemical warfare, particularly as regards the use of chemical agents against flora. The UN General Assembly last addressed this matter in 1969 when it adopted resolution 2603 A (XXIV) by a vote of 80 to 3, with 36 abstentions, declaring as contrary to the generally recognized rules of international law the use in international armed conflicts of any chemical agents of warfare which might be employed because of their direct toxic effects on man, animals or plants. Finally, it is to be recalled that Article 23 (a) of the 1907 Hague Regulations respecting the Laws and Customs of War on Land[17] forbids the employment of poison or poisoned weapons; this rule has customarily been interpreted as prohibiting, *inter alia*, the poising of wells and other water supplies.

With regard to the future, three current activities of the UN Organization should be noted which relate to war and which have, or could have, direct implications for environmental protection. The first are the endeavours being made to secure agreement on further disarmament and related measures. Two sets of discussions here are particularly relevant: those having to do with the complete prohibition of all nuclear-weapons tests, and those seeking an international convention banning the development, production and stockpiling of chemical weapons. Both endeavours have been proceeding, in one forum or another, for several years.[18] States party to the 1972 Biological and Toxin Weapons Convention have committed themselves, under Article IX, to continue negotiating in good faith for a corresponding agreement on chemical weapons. Second, and also being conducted within the UN disarmament machinery, is the UN Conference on Prohibitions or Restrictions of Use of Certain Conventional Weapons Which May Be Deemed to Be Excessively Injurious or to Have Indiscriminate Effects. The preparatory conference commenced in August 1978, the conference itself being scheduled for

---

[15] *Supra*, note 9, at pp. 114-118.
[16] League of Nations, *Treaty Series*, Vol. 94 (1929), No. 2138.
[17] A/9215.
[18] For a review, see *The United Nations Disarmament Yearbook*, volume 2:1977 (New York: United Nations, Sales No. E.78.IX.4, 1978), at pp. 95-108 and 193-210.

September 1979. If specific "conventional weapons" exist which are capable of assaulting entire human communities through environmental damage, their use in such a manner would indisputably have "indiscriminate effects". Thirdly, there are the attempts being made to cope with the problems presented by what are called, in General Assembly resolution 3435 (XXX), "material remnants of war". That resolution requested the Governing Council of UNEP to examine the problems further, and two reports have thus far been made to the General Assembly.[19] UNEP is currently charged with the task of carrying out and promoting, in collaboration with appropriate components of the UN system, studies on the environmental effects of these remnants, particularly mines.[20]

APPROACH

It is not the intention in this study to provide a compendium of the forms which enviromental damage from weapons might take. All features of the natural environment are vulnerable to one type of weapon or another if used on a sufficiently large scale. Forests may be destroyed, wildlife killed, vegetation stripped, soil shattered, watercourses broken, artefacts destroyed: the list could be extended indefinitely, and in the end would amount to no more than a statement of the obvious, embedded in descriptive detail of uneven interest and wider significance. The potential human consequences could not easily be conveyed by such an approach, so that in comparison with the direct effects of weapons on people and human societies such a study of the environmental effects would appear trivial, even demeaning. Yet the fact remains that there are numerous examples in history, both recent and ancient, of human societies suffering appallingly from military assaults on their natural environment. As weapons technology moves towards greater and greater destructive potential within a world of sharpening conflict, the probability of such episodes recurring must increase. If the probability and the dangers are to be assessed so that precautionary measures may be taken as necessary, it is not descriptive study of the environmental impacts of weapons that is needed, but analytical study. The basic requirement is for a conceptual framework which can be used for assimilating the empirical knowledge that already

[19] A/31/210 and A/32/137.
[20] UNEP Governing Council decision 6/15 of 15 May 1978

exists, for revealing crucial ignorance, and for ranking varieties of damage and risk.

The present study is a very preliminary attempt at providing such a framework. The approach used rests on the proposition that no living organism, whether man, microbe or any other, is self-sufficient. All depend for survival upon their environment, which, most basically, is a source of nutriment and a sink for potentially lethal waste products. Since the resources of the environment of an organism are finite in both these respects, survival of an individual organism is also dependent upon the density of the population of which it forms a part. It is further dependent upon the community of populations of other categories of organism of which its own population forms a part, for different categories of organism place different demands upon their environment, and those of one category may conflict with those of another; equally, the life processes of some categories of organism may act to provide sustenance in the environments of other categories. The community as a whole cannot exist without the cycling of materials and the flow of energy through it. This leads to a broad conception in which human societies are seen as being linked to one another and to their natural environment within an enormously complex web of interactions and interdependencies: a fabric whose integrity might be jeopardized by damage at any one point, as by the destructive action of weapons. Ecologists have coined the term "ecosystem" to convey such notions of interaction and interdependence. A study directed more upon the broad ecosystemic effects of weapons than upon specific effects on individual ecosystem components should render the full risks of environmental damage more readily comprehensible.

That said, a fundamental limitation of the approach must also be noted. There exists a considerable body of knowledge about particular aspects of the behaviour of particular ecosystems under normal circumstances. But in many, perhaps most, cases this is insufficient to predict ecosystem behaviour under abnormal circumstances. Therefore, in seeking to explain how ecosystems are likely to respond over time to the manifold insult that weapons may impose, this study can frequently deal only in ill-defined probabilities and unproven hypotheses. To the scientist operating at the borders of knowledge, this is a familiar enough situation. But it is not one that lends itself easily to the formulation of policy or the mapping out of priorities for action within the international community. Yet it seems clear that the dangers are potentially of such a magnitude that action cannot prudently be allowed to wait upon certainty.

Uncertainty prevails in many fields of human enterprise, and various ways have been developed for accommodating it. Above all, there are techniques of systems analysis. This is a further reason why the present study used a systems approach. The assumptions which it imputes need to be made explicit here. The most fundamental is that it makes sense, not only in abstract intellectual terms, but also in concrete practical terms, to compartmentalize perceptions of the living world into discrete biosystems, of which ecosystems are one category. Some such simplification is needed if order is to be discerned and behaviour thereby explained or predicted; but over-simplification may distort understanding. Above all, the boundaries of each such system must somehow be drawn so as not to exclude factors which may, in some as yet unknown way, govern broad system behaviour, but without at the same time imposing an intellectually unmanageable degree of complexity. Acceptance of the concept of ecosystem implies that such compromises are both feasible and useful.

To focus on ecosystems is, in conceptual terms, to limit attention to a broad band within the spectrum of biological organization across which weapons may exert their effects. How is that band to be marked out? The problems alluded to in the previous paragraph are, in systems analysis, routine matters of aggregation subject to the well-established principle of integrative levels. Prediction of the behaviour of a system does not require precise understanding of how each component is structured from simpler subcomponents. Biological organization can be conceived hierarchically (for example: gene, cell, organ, organism, population, community) and a biosystem, meaning a set of interdependent biotic and abiotic components interacting with one another, can be defined at each hierarchical level. "Ecosystems" are biosystems defined at the community level. They can be portrayed in different ways, with greater or less representation of lower hierarchical levels as subcomponents. Their utility, like that of any conceptual model, depends upon the degree to which their structure, including their hierarchical breakdown, fits the process they are set to elucidate and the behaviour they are set to forecast, in the present case predicting ecological sequelae at the community level of different types of weapons-employment. Clearly, results will be unobtainable if the ecosystem models employed as the predictive tool lack sufficient definition to portray sensitivities to different types of weapons-effect. This posits a minimum degree of hierarchical breakdown: a minimum complexity of model structure: a minimum number of factors to be taken into account. For example, chemical weapons are, by definition, ones that act through toxicity, lethal or nonlethal;

if their effects on ecosystems are to be assessed, and differentiated from other forms of insult, the ecosystem model used must include subcomponents that are sensitive to toxicity. However, if realism is to be sustained, other subcomponents of the model may need to be portrayed at an equivalent degree of hierarchical breakdown so as to capture, for instance, the operation of possible repair mechanisms. The question then is whether such complexity is manageable and, more particularly, whether such complexity is manageable and, more particularly, whether the available data base is sufficiently extensive and detailed.

It is clear from these considerations that a rigorous application of quantitative systems analysis would require more time and resources than are available for the present study. It is to be doubted, moreover, whether there is yet sufficient empirical knowledge to draw upon. The alternative is to rely instead on a qualitative treatment. Systems-analysis techniques are now becoming available, most notably that of structural modelling,[21] which, when applied to uncalibrated models of ecosystems, might reveal much about likely weapons-effects at present obscured by ignorance and uncertainty. However, even these techniques, which require appreciable computer time, are beyond available resources. Instead, a much cruder approach is used. It is designed to give an approximate idea of the full dimensions of the problems and can be taken only as a preliminary mapping-out exercise, useful as a point of departure for the more solid investigation that is clearly necessary.

The approach employs a simple conceptual model of ecosystem vulnerability, which is to say the sensitivity of an ecosystem to disruption following perturbations originating in the action of weapons. Chapter 2 sets out a classification of weapons of mass destruction. In use, each such class of weapon has its own characteristic set of weapons-effects, and an ecosystem can be expected to display varying degrees of sensitivity towards each such effect. Available information on the relevant effects of each class of weapon is reviewed in turn. Chapter 3 sets out a classification of ecosystems and, for the classes of ecosystem most pertinent to the present study, reviews in turn what is known about possible major sensitivities to perturbations of the type that weapons might cause. The ecosystem-classification used is necessarily a

---

[21] See, in particular, M. McLean, P. Shepherd and R. Curnow. "Techniques for analysis of system structure", *SPRU Occasional Paper Series* (Science Policy Research Unit, University of Sussex) No. 1, February 1976.

coarse one; a finer classification would clearly be needed in any further study. Chapter 4 reviews the very limited amount of empirical information available on ecosystemic effects of weapons, namely data from ecological surveys of actual weapon test-sites and theatres of war. This is used to develop the theoretical analysis of ecosystemic effects of mass-destruction weapons that is obtainable by relating the weapons-effect characteristics set out in Chapter 2 to the ecosystem-sensitivity characteristcs set out in Chapter 3, having regard to the possible scales on which the weapons might be used. This analysis is then used to explore the possible consequences for human populations that are components of, or otherwise sustained by, ecosystems perturbed by weapons. Conclusions are drawn regarding action at the international level.

Chapter 2

WEAPONS

CLASSIFICATION OF WEAPONS

A weapon is a device for damaging a target in a manner that is predictable enough for military purposes. The many varieties that have been developed over the millenia may conveniently be classified here according to the type of damage-agent that is used. Table 1 sets out such a classification. The broad characteristics of each class are as follows.

Piercing weapons are the archetypal weapon of human combat. The firearm projecting a small, pointed, solid missile at high velocity is the dominant embodiment today of a line of development that stretches back through the bow and arrow, the javelin, the spear, the sword, the knife and the axe. Since such weapons are unlikely to have a discernible effect on ecosystems as extensive as the ones considered in this study unless used on an implausibly large scale, they will not be discussed further.

High-explosive weapons are designed to cause physical damage by means of intense pulses of energy released from chemical compositions made to undergo extremely fast combustion reactions. The energy may be transmitted to the target in the form of a shock-wave (blast) or in the form of high velocity fragments of a material encasing the explosive composition.

Incendiary weapons are those that are primarily designed to set fire to objects, or to cause burn injury to living organisms, through the action of flame and/or heat produced by chemical reaction of a substance delivered on the target. Incendiary weapons using gelled formulations of hydrocarbons (such as napalm) or of hydrocarbon-derivatives are sometimes subcategorized as "flame weapons".

TABLE 1. CLASSIFICATION OF WEAPONS

Class of weapon	Damage Agent
Piercing	Penetrators (bullets, bayonets, etc.)
High-explosive	Blast
	High-velocity fragments
Incendiary	Flame
	Heat flux
Chemical	Toxic chemicals
Biological	Pathogenic microbes or toxins
Radiological	Radioactive substances
Nuclear	Blast
	Thermal radiation
	Ionizing radiation
	Radiotoxic substances

Incendiary weapons, together with high-explosive and piercing weapons, constitute a category commonly regarded as "conventional weapons". However, some of the newer types of incendiary weapon, such as the "controlled chemical fireball" devices now beginning to supersede napalm, and of explosive weapon, such as fuel-air explosives, are capable of exerting effects so gross in their destructiveness, and so novel in their character, that "conventional" appears as a remarkable, even pernicious, misnomer.

Chemical weapons depend for their effects upon chemical-warfare agents, these having been defined in a previous report of the UN Secretary-General[22] as "chemical substances, whether gaseous, liquid or solid, which might be employed because of their direct toxic effects on man, animals and plants". Several different types of chemical-warfare agent have been developed for use. In some, toxicity is exploited as a means for securing only transient target effects, such as a temporary incapacitation or harassment of enemy troops or premature leaf-abscission in deciduous trees; in others, it is exploited as a lethal mechanism.

Biological weapons are those which depend for their effects on what are termed in the 1972 Biological Weapons Convention (see page 5) as "microbial or other biological agents, or toxins whatever their origin or method of

---

[22]*Chemical and Bacteriological (Biological) Weapons and the Effects of Their Possible Use*, A/7575/Rev.1 (UN Sales No. E.69.I.24).

production, of types and in quantities that have no justification for prophylactic, protective or other peaceful purposes". Such agents of biological warfare may comprise any sort of pathogenic microorganism or toxic material generated by living organisms.

Radiological weapons resemble chemical weapons except that the agents which they would use be chosen for their radioactivity, and hence their radiation or radiotoxic effects, rather than their chemical toxicity. Conceivable radiological-warfare agents include spent fuel from nuclear power reactors; materials such as tantalum which can readily be converted into powerful gamma-ray sources by neutron activation; and intensely radiotoxic alpha-emitters, such as plutonium.

Radiological weapons may be classed with chemical and biological weapons, and differentiated from other categories of weapon, in the biospecificity of their action: "CBR" weapons are designed to damage or destroy only living organisms. Weapons lacking biospecificity function by non-discriminating physical, rather than biological, actions and may destroy animate and inanimate targets alike.

Nuclear weapons, whose actions derive from controlled thermonuclear fusion and/or nuclear fission chain-reactions, are in effect composite incendiary, explosive and radiological weapons of enormous power. They can be tailored in design or in mode of employment so that one or other of the three main effects predominate, though all will contribute to the overall damage caused.

In common usage, the expression "weapons of mass destruction" is the antonym to that of "conventional weapons" noted on page 12. In fact, the amount of destruction which a particular class of weapon is capable of causing is a function not only of inherent characteristics but also of the scale on which the weapons are used and the vulnerability of their targets. Thus, while it is the case that a single nuclear weapon, for example, is capable of causing enormous destruction against virtually any type of target, so too are conventional explosive weapons if used in, say, prolonged artillery or aerial bombardments, or conventional incendiary weapons if used against readily combustible targets.

Nevertheless, this usage has over the years acquired a more precise meaning than its actual wording suggests; and in international disarmament negotiations it is essentially a technical term. Here, the expression "weapons of mass destruction" imports the general tenor of a resolution adopted in September 1947 by the Working Committee (consisting of States then members of the Security

Council) of the UN Commission for Conventional Armaments, by a vote of 7-2-2; the resolution defined the armaments that lay outside the terms of reference of the Commission as "weapons of mass destruction" including "atomic explosive weapons, radioactive material weapons, lethal chemical and biological weapons, and any weapons developed in the future which have characteristics comparable in destructive effect to those of the atomic bomb or other weapons mentioned above." The Commission approved this definition in August 1948, with two opposing votes.[23] Since then, the expression has been used repeatedly in a variety of intergovernmental fora, though with no further attempt at definition. It is employed without gloss in both the 1967 Outer Space Treaty and the 1971 Seabed Treaty (see page 4).

The present study uses a less restrictive interpretation of "weapons of mass destruction". The underlying concern is the possibility that use of weapons might cause serious damage, especially desertification, in major ecosystems. This is itself as good a way as any other of defining weapons of mass destruction. The remainder of this chapter is therefore given over to a more detailed discussion in turn of each of the classes of weapon that might have such effects.

It should be observed, first, that while there are several different classes of weapon that need to be considered, different likelihoods of use attach to each one. This modifies the relative degree of concern warranted by each class, and requires that more attention be given to some than to others. Three criteria would seem to be important in this respect: military utility, legality and availability.

The practice in wartime of damaging or impeding an enemy by ecosystem-ically-mediated rather than direct weapons-effects is not new. "Scorched earth" tactics, for example, have ancient precedents. For the most part, however, this form of warfare has been practised by combatant forces on the defensive or facing defeat. Its military utility as an offensive tactic seems to be limited by several factors, the net effect of which is to make it generally more profitable for an attacking force to use its weapons directly against the opposing forces rather than against their environment. Guerrilla warfare is a major exception, for here one side deliberately exploits its environment for protection, for sustenance, for surprise, and generally as a means for avoiding direct military engagement by opposing forces. The latter may therefore be strongly

---

[23] S/C.3/SR.13.

tempted, for military reasons alone, to use ecosystemically-mediated modes of offensive action. Recent wars in several parts of the world, particularly in tropical regions, have seen such action taking place. But in no case has its efficacy, analysed retrospectively using purely military criteria, been conclusively demonstrated. What has been demonstrated is that the deleterious side effects of such operations may be very great: a factor countervailing whatever military benefits might or might not accrue. The weapons used in such operations have commonly been acquired specially for that purpose, most notably ones based on chemical herbicides. This suggests that, on the criterion of military utility, the weapons of gravest concern as regards ecosystem mass-destruction are a narrow range of special-purpose weapons. Weapons outside this range would seem to warrant attention here only if, in their normal mode of use, they are likely to cause unavoidable ecosystemic damage.

Regardless of the purposes for which they are used, the military utility of some classes of weapon will be limited by their very nature. In particular, radiological weapons have many operational limitations which can scarcely fail to reduce their military attractions. Thus, gamma-active radiological-warfare agents would necessitate complicated shielding arrangements if their users were not to be their first victims; and both these and the few candidate beta-active agents decay sufficiently fast to preclude stockpiling. The alpha-active candidate agents do not have these limitations, but since their radio-toxicity may take years or decades to become significantly manifest, their military applications would seem to be few. In short, radiological weapons do not seem to make much military sense, though exceptional geographic or demographic circumstances can perhaps be envisaged in which they might. Although if they were used they might well cause exceedingly grave ecosystemic damage, they will not be considered further in this study.

To the extent that a category of weapon is difficult to acquire, or illegal to use or possess, the various threats which it poses are somewhat lessened. This is not a matter that bears developing here. It is worth noting, however, that, at the present state of their technology, biological weapons (whose possession is in any case illegal) may well be almost as difficultly accessible, in militarily-attractive forms and militarily-significant quantities, as nuclear weapons. Nor, by and large, are the more potent types of chemical weapon easy to acquire, involving as they do advanced chemical production technology. Herbicidal chemical weapons are an exception, for several of the herbicides which are manufactured and marketed on a large scale for agricultural and forestry purposes are readily adaptable to weapons.

The following paragraphs concentrate on those weapons which, in the light of the foregoing considerations, may be taken as posing a particularly grave threat to ecosystems. The treatment is representative rather than comprehensive. The main purpose is to identify those particular effects of each type of weapon to which certain ecosystems might be especially sensitive.

HIGH-EXPLOSIVE WEAPONS

High-explosive weapons can be tailored so that one or the other of their damage-agents dominates their effects, the result being predominantly either a blast weapon or a fragmentation weapon. More commonly, the weapons are used in a "general purpose" form in which both actions are strongly displayed. Any of the three varieties of high-explosive weapon—blast, fragmentation or general-purpose—may cause damage in both the biomass and the geomass of an ecosystem. The blast may cause local disturbance of the soil by cratering. Its biocidal action will extend radially outwards in all directions from the point of burst. The intensity, and hence the lethality, of the blast diminishes rapidly with distance, however, and physical features of the environment may afford some shielding. Much the same can be said of the fragmentation effects, though the uniformity of a blast wave is such that, over unshielded areas, no single point will escape its effects, so long as its intensity remains above the threshold of effectiveness. The blast effects of high-explosive weapons are likely to be the dominant stress factor in whatever ecosystemic impact the weapons may have. The fragmentation effects could be more severe in some ecosystems than in others, especially in forests where fragments implanted in trees could open a way to invasion by micro-organisms.

Figures are available which allow rough estimates to be made of total areas likely to be affected in particular respects by a given weight of high-explosive attack. With regard to general-purpose bombs and artillery shell, Westing[24] has estimated that, on average, an area of about 12.5 m^2 might be exposed to environmentally significant blast and fragmentation damage per kilogram of munitions expended; people in such an area would face a high probability of death, and woody vegetation would be effectively demolished. The cratering

---

[24] A. H. Westing, for the Stockholm International Peace Research Institute, *Ecological Consequences of the Second Indochina War* (Stockholm: Almqvist & Wicksell, 1976) p. 22.

effect of a 250-kg general purpose bomb may displace, on average, about 70 m^3 of soil, though this will vary within wide limits according to the type of soil and the manner in which the bomb is fused.

If it is correct to identify blast rather than fragmentation as the factor of greatest ecosystemic significance among the effects of high-explosive weapons, it follows that the high-explosive weapons of greatest concern here are the blast-maximized ones. The recently introduced fuel-air explosive munitions fall within this category. Such weapons dispense a pancake-shaped cloud of aerosolized fuel low over the target area, which, after a short delay, becomes diluted with air to the point where it is explosive; the weapon then injects detonators into the cloud. The uniformly distributed, wide-area blast effects of such weapons can reach inside enclosures and into places that would be shielded against conventional forms of high-explosive attack. A much larger proportion of the biomass would thereby be subjected to biocidal effects. In their primitive forms, fuel-air explosive weapons were capable of stripping vegetation and blowing down trees over an area of about 10 m^2 per kg of fuel; their efficiency has increased greatly during the past decade of research and development.

INCENDIARY WEAPONS

Though incendiary weapons are not all designed for the purpose of initiating self-propagating fires within a target area, all of the many different types that have been developed are capable of doing so if the target is sufficiently inflammable. It is in the possibility of such fires being created and then spreading across rural areas that incendiary weapons are of greatest concern here. Particular attention must attach to flame weapons, as defined on page 11, for these are the subcategory of incendiary weapon likely to cause the most extensive damage in a rural environment when measured in terms of average area affected for a given weight of weapons, even if no self-propagating conflagration results. Napalm firebombs may occlude within the fireballs which they create an area of about 6 m^2 per kg of napalm. How severely that area is damaged will depend upon its nature; actual operational experience with the older forms of napalm firebomb indicates that, in tropical deciduous forest, complete destruction of vegetation is to be expected within the fireball area of three such firebombs falling together. The latest forms of flame weapon, those that disseminate thickened forms of such high-energy pyrophoric chemicals as triethylaluminium, are very much more destructive, primarily because a higher proportion of their heat output is in the form of thermal radiation. They thus have a substantially greater likelihood of initiating spreading fires.

The factors that govern the ease with which fire can take hold and then spread have been closely studied. As a result, the types of rural area that are most at risk can be specified rather well. By and large they are the areas where the natural incidence of wildland fires is highest. Thus, tropical savannah is particularly vulnerable during the dry season, whereas tropical rain-forest is scarcely vulnerable at all. Once they take hold, wildland fires can become very large and may extend for hundreds, or even thousands, of square kilometres. They may take a long time to burn out. Forest fires have been known to smoulder on all winter under a blanket of snow, becoming active again the following summer once the fuels dry.[25]

The ecosystemic effects of fire can be expected to reside in the universally biocidal action of strong heat flux, and, perhaps to a lesser extent, in physical changes brought about in the soil should it become scorched. It is worth noting that ecosystems which commonly experience natural wildland fires are likely to be more robust than others in the face of fires initiated by incendiary weapons. However, in that forest fires are known to favour, and sometimes be required for, the germination of previously dormant seeds of certain species (e.g. *Adenostoma fasciculatum* in California), even mild fires could grossly alter the composition of a forest community.

CHEMICAL WEAPONS

Chemical weapons, in the sense mentioned on page 12, have been used in large quantities in two wars only. About 125,000 tons of chemical agent were employed during World War I, and about 96,000 tons during the Viet-Nam conflict. There have been a few other instances of chemical warfare this century, but on a much smaller scale.

The stress that chemical weapons may place on ecosystems resides in the biospecificity of their toxic action. This will immediately endanger at least some of the living organisms with which the chemical-warfare agent comes into

---

[25]*Napalm and Other Incendiary Weapons and All Aspects of Their Possible Use*. Report of the UN Secretary-General, A/8803/Rev.1 (Sales No. E.73.I.13), at para. 75.

contact after dissemination. There will also be a residual longer-term threat whose magnitude will vary greatly from one type of chemical to another and from one environment to another. The determinants of the residual threat may be grouped into two broad factors: persistence and mobility. Persistence is a measure of the time period during which the chemical remains present and active in the soil and biota. Mobility refers to the capacity of the chemical for entering and moving up food-chains. The latter is a process during which a chemical may sometimes become concentrated, depending upon the degree to which it is retained in living tissue, thereby presenting an increased toxic hazard at higher trophic levels. Both intrinsic and extrinsic factors contribute to persistence: on the one hand, the volatility of the chemical, its capacity for adsorption or absorption, and its chemical stability towards moisture, ambient heat stress, microbial attack, etc.; on the other hand, factors such as rainfall, ambient temperature regime and soil type. It is to be noted, however, that while the factors which impart environmental persistence and mobility to a chemical can be specified rather precisely, it is rare for the chemical and physical properties of individual chemicals, including chemical warfare agents, to be known sufficiently fully for these factors to be adequately assessable.[26]

The chemical-warfare agents used in World War I were for the most part lethal anti-personnel chemicals. Though they caused more than a million human casualties, their longer-term ecological consequences appear to have been slight (see page 43). The principle new class of agents within this category to have emerged since World War I comprises the organophosphorus anti-cholinesterases known as "nerve gases": volatile or involatile liquids which are estimated to be lethal to man at milligram dose levels. They are one to two orders of magnitude more toxic than their predecessors, and might for this reason have a considerably greater ecological impact. It seems clear that if nerve gases were used at levels lethal to human beings (application densities upwards of 0.5 kg per hectare), they would probably destroy a high proportion of whatever nonhuman vertebrate population was thereby also exposed, and many of the invertebrates as well, particularly arthropods. Vegetation would

---

[26] In the face of these and other deficiencies in knowledge relating to the environmental impacts of individual chemicals, attempts are being made to devise simple early-warning indications on the basis of whatever data are available; see, particularly, R. C. Harriss, "Suggestions for the development of a hazard evaluation procedure for potentially toxic chemicals", Monitoring and Assessment Research Centre (Chelsea College, University of London) research memorandum, MARC Report No. 3 (1976).

probably mostly be spared, though, as Worthley[27] and Howells and Hambrook[28] have shown, the nerve gases are not devoid of phytotoxicity. The contamination of vegetation would, however, present a serious hazard to herbivores feeding on it for some while afterwards, possibility for several weeks in the case of the involatile nerve gases.

When taken into solution, the nerve gases hydrolyze quite rapidly under environmental conditions, and are metabolized by plant species, to yield alkylphosphonates. The latter tend to persist as such, but so far as is known, they do not present any particular toxic hazard to living organisms. All in all, problems of environmental persistence and food-chain concentration, such as are well-known with, for example, some organochlorine pesticides, seem unlikely to be encountered with the nerve gases. Though quantitative data do not appear to be available, it would seem that the persistence of a nerve-gas hazard in ecosystems is unlikely to have a half-life exceeding 2-3 months.[29]

It is conceivable that synthetic chemicals of high toxicity in man, lethal or incapacitating, might emerge as candidate chemical warfare agents in place of or alongside the nerve gases; and that such new antipersonnel agents might present an ecological threat exceeding that of the nerve gases. Chemicals such as 2,3,7,8-tetrachlorodibenzo-p-dioxin (TCDD) lend credence to this possibiliy. The episode at Seveso in Northern Italy in July 1976, for example, when an estimated 2.5 kg of TCDD was distributed as a result of an industrial accident over some hundreds of hectares of semirural terrain was an ecological catastrophe on any count; yet the release was trivial in size in comparison with what would have to be expected during actual chemical warfare. TCDD has a lethal toxicity in mammals comparable with that of the nerve gases, though it is far slower to take effect. and at sublethal dosages it has a wide range of chronic toxicological manifestations, including teratogenicity. Its chemical

---

[27] E. G. Worthley, "The toxicity of VX to various plants", U.S. Army Edgewood Arsenal technical report No. EATR 100-9, June 1970.
[28] D. J. Howells and J. L. Hambrook, "The phytotoxicity of some methylphosphonofluoridates", in *Pesticide Science*, 3: 351-356 (1972).
[29] The following references provide a point of access into the literature on the environmental persistence of the nerve gases, and also include data bearing upon the threat which these agents may pose to water supplies: J. Epstein, "Rate of decomposition of GB in seawater", *Science* 170: 1396-1398 (1970); H. O. Michel *et al.*, "Detection and estimation of isopropyl methylphosphonofluoridate and O-ethyl S-diisopropylaminoethyl methylphosphonothioate in seawater in parts-per-trillion level", *Enviromental Science and Technology* 7: 1045-1049 (1973); and J. Epstein, "Properties of GB in water", *Journal of the AWWA*, January 1974, pp. 31-37.

and physical properties are such that it persists and has some mobility in ecosystems: half-lives of the order of years have been observed in different instances. There is evidence that it is capable of entering and moving up certain food-chains that culminate in man. Ramel[30] has provided the most recent review of these and other properties of TCDD.

The chemical agents employed during the Viet-Nam conflict were for the most part herbicides used by contact application to defoliate or otherwise remove forest cover and to destroy certain food-plant cultivations. The results of a number of preliminary assessments of the overall ecological consequences of these operations have been published (see pages 43 and 44). The herbicides employed were commercial formulations, but the application densities at which they were used (5-30 kg of active ingredient per hectare) substantially exceeded those of agricultural or silvicultural practice. The greater part comprised formulations of 2,4-dichloro- and 2,4,5-trichloro-phenoxyacetic acid derivatives, though substantial quantities of cacodylic acid and picloram were also used.

The selective biospecificity of herbicides is the exact opposite to that of the nerve gases, toxicity towards plant species far exceeding that towards animals. Their impact on ecosystems is therefore likely to be substantially different: more radical, because of their selective assault at the first trophic level, and hence potentially far more severe. Their civil applications are such that a great deal of research has been, and continues to be, devoted to species-selectivity. The result is that a broad range of different chemicals is available for use against different types of plant species. Similar variety is also available as regards such factors as environmental persistence. Herbicides may therefore present a potential for environment degradation nowhere near approached by their past military applications.

In addition to direct phytotoxicity, there are other effects for which herbicides might be chosen for use in chemical weapons. For example, the herbicides that are most effective in stripping forest vegetation appear to exert relatively little effect on the microflora of the soil. and those that have been in weapons have had relatively short persistence in the soil. Other agents differing in either or both of these respects might conceivably prove attractive

---

[30]C. Ramel (ed.), *Chlorinated Phenoxy Acids and Their Dioxins: Mode of Action, Health Risks and Environmental Effects* (Swedish National Science Research Council, Ecological Bulletin No. 27, 1978).

in longer-term strategies of food-denial via soil sterilization. Such applications would greatly compound the overall insult to the ecosystem.

BIOLOGICAL WEAPONS

Biological weapons, as defined on page 12, have never been used on a major scale in war, though as with chemical weapons there have been many allegations and instances of minor use. It is not easy, therefore, to foresee the most likely forms which such weapons might take if they were to be used in the future, still less the possible ecological consequences. Moreover, the rapidity with which microbiological science is advancing, as in the current opening up of the manifold possibilities of recombinant DNA, further increases the uncertainty. With the entry into force of the 1972 Biological Weapons Convention, the development programmes that had been under way became illegal; and there are strong indications that, despite the many years of effort that had gone into the development of biological weapons much more still remained to be done at both the technical and the military-doctrinal level before militarily attractive weapons could have resulted. Some production and stockpiling of biological weapons had, however, taken place, including anticrop weapons designed to initiate epiphytotics within cereal cultivations. Though defunct, these weapons provide the only solid guidance for present purposes.

With the exception of the toxins, biological-warfare agents are living organisms, each with its own particular preferences as regards habitat, nutrients, etc. The further the conditions of their employment for biological-warfare purposes diverge from their natural conditions, the less likely they are to survive and cause disease. For large-scale effects, the agents would have to be disseminated in the airborne, aerosolized form. It seems that most pathogens are incapable of withstanding the unnatural stresses and deprivations which this would entail.

But there are some pathogens which undoubtedly do have this capacity, or which can be bred in amenable mutant forms or artificially protected; and for them it is possible to envisage disease being initiated by, for example, a small fleet of aerosol-generating aircraft over an area of tens, hundreds, even thousands of square kilometres. The consequences of such attacks would hinge on the ability of the micro-organisms, in their high stressed condition, to establish themselves within the new environment into which they were deposited. So much would depend upon circumstances that little is possible in the way of prediction either of the magnitude of the resultant target effects or of ecological sequelae.

As to the latter, the dangers presented by some pathogens are clearly greater than others. For example, the causative agent of anthrax, a frequently fatal disease that can strike most mammals and a variety of other animals, is a bacterium that can sporulate; its spores can remain alive in soil for decades under a wide range of climatic conditions. Some of the candidate viral agents might rather easily establish themselves within the cells of other living organisms, in which event the latter could constitute persistent reservoirs of the disease if they did not themselves succumb to it. Insect populations feeding off such carriers could become vectors of the disease. Related to this form of potential environmental persistence is a third form, in which pathogens capable of doing so initiate contagious disease. It appears to have been the case, however, that, with the important exception of antiplant biological weapons, contagious-disease agents were not favoured in the later stages of those of the biological-weapons development programmes about which details are publicly available, though they were favoured in the initial stages. The unpredictabilities in the dynamics of contagious disease are such that weapons which might initiate spreading diseases will be even more deficient than other types of biological weapon as regards the military requirement for controllability of effect.

Toxins are known that are up to four or five orders of magnitude more toxic to man than the most potent of the synthetic poisons for which antipersonnel chemical weapons have been designed. It is therefore possible to envisage their use in extraordinarily destructive weapons comparable in area-effectiveness with those based on infective pathogens. So far as is known, none of the toxins that have been developed as fills for wide-area biological weapons are stable in the environment for periods exceeding a few days; and their toxicity varies very widely among different species, none being notably toxic to plants. Moreover, unless substances such as hydrogen cyanide or fluoracetic acid are considered to be toxins, they are unlikely to be available in sufficient quantities for large-scale use. By and large, therefore, the ecosystemic impact of toxin weapons is unlikely to exceed that of, say, nerve-gas weapons.

As with chemical weapons, it is in their biocidal action that biological weapons may seriously stress ecosystems. But in their biospecificity, biological weapons will generally be more species-selective than chemical weapons, so much so that their overall ecosystemic impact may be very different, especially if the agents establish themselves as persistent reservoirs of plant or animal disease.

NUCLEAR WEAPONS

The vastness of the energy-release from nuclear weapons, and hence their destructive potential, is difficult to comprehend. The yield of a single moderate-sized hydrogen bomb may exceed the total energy of all the explosives used during both World Wars. The energy release would be in several forms. Those of greatest relevance here are, in descending order of magnitude, blast, thermal radiation, and nuclear radiation. In different ways each is capable of placing enormous stress upon ecosystems in which the weapons are employed. The stress may be direct, as in the physical destruction due to blast or fire, or in the biospecific damage due to ionizing radiation or radiotoxicity. Or the stress may be indirect: a secondary consequence of atmospheric or geospheric disturbances caused by the weapons which may lead to weather or even climate modification.

As to the direct effects of nuclear weapons, those originating in blast and thermal radiation differ broadly only in degree from what has already been described in connexion with high-explosive and incendiary weapons. But those originating in nuclear radiation are unique among weapons. Living organisms vary greatly in radiosensitivity. Some insect species may be able to survive radiation doses many hundreds of times greater than those that would rapidly kill the more sensitive vertebrate animals, such as man. Likewise, among plant species, some organisms may be three orders of magnitude more sensitive than others, the order of decreasing radiosensitivity tending to be trees, shrubs, herbs and thallophytes. Radiation may have overt somatic effects leading to death or growth inhibition or a reduction in reproductive capacity; and there may also be a wide range of genetic effects following upon increased mutation rates. Whicker and Fraley[31] have reviewed these effects exhaustively and in detail with regard to plant communities. Table 2 illustrates the relative magnitudes of the three major weapon-effects in their action upon particular components of terrestrial ecosystems.

---

[31] F. W. Whicker and L. Fraley, Jr. "Effects of ionizing radiation on terrestrial plant communities", *Advances in Radiation Biology* **4**: 317–366 (1974).

TABLE 2. EFFECTS OF GROUND-BURST NUCLEAR WEAPONS ON DIFFERENT COMPONENTS OF ECOSYSTEMS

Type of damage	Area over which damage may occur (*hectares*)	
	20 kt atom bomb	10 Mt hydrogen bomb
Craterisation by blast wave[a]	1	57
Vertebrates killed by blast wave[b]	24	1 540
All vegetation killed by nuclear radiation[c]	43	12 100
Trees killed by nuclear radiation[d]	148	63 800
Trees blown down by blast wave[e]	362	52 500
Vertebrates killed by nuclear radiation[f]	674	177 000
Dry vegetation ignited by thermal radiation[g]	749	117 000
Vertebrates killed by thermal radiation[h]	1 000	150 000

Source: A. H. Westing (for the Stockholm International Peace Research Institute), *Weapons of Mass Destruction and the Environment* (London: Taylor & Francis, 1977), page 17.

[a] Refers to dry soil. A subsurface burst could craterize four times as large an area as a surface burst. Nuclear warheads exploding above the surface would produce no craters at all if the burst were sufficiently high, but the nuclear and thermal radiation effects would then extend over larger areas.

[b] Refers to areas over which the transient overpressure, ignoring Mach reflection, would be likely to exceed 345 kilopascal, this being the overpressure for about 50 percent lethality among large mammals, including man.

[c] Refers to areas over which the early radiation dose would be likely to exceed 70 Kilorad.

[d] Refers to areas over which the early radiation dose would be likely to exceed 10 kilorad.

[e] Refers to areas over which the transient wind velocity at the shock front, ignoring Mach reflection, would be likely to exceed about 60 metres per second. Such a wind would be likely to blow down about 90 percent of the trees in an average coniferous forest or a deciduous forest in leaf.

[f] Refers to areas over which the early radiation dose would be likely to exceed 2 kilorad.

[g] Refers to areas over which the incident thermal radiation would be likely to exceed 500 kilojoules per square metre for the atom bomb or 1,000 kJ/m^2 for the hydrogen bomb, if the weapons were detonated on a clear day having a visibility of 80 km.

[h] Refers to areas over which, on a clear day of 80 km visibility, the incident thermal radiation would be likely to exceed that which would have a 50 percent lethality for exposed pigs (380 kJ/m^2 for the atom bomb and 750 kJ/m^2 for the hydrogen bomb).

The data given in Table 2 on the effects of nuclear radiation relate only to so-called "early radiation", that is to say, to the radition dosage during the 24-hour period immediately succeding the explosion. Of the total energy released by a nuclear explosion that is in the form of nuclear radiation, 10–15 percent will remain undissipated at the end of that 24-hour period. This residual radioactivity will constitute a chronic stress upon ecosystems in which it is deposited (as radioactive fallout). Of particular significance are the radionuclides caesium-137, strontium-90, carbon-14, hydrogen-3 (tritium) and iron-55. Apart from the carbon-14, they will concentrate predominantly in the soil of terrestrial ecosystems; some fraction of them will be available for uptake by plants and subsequent transfer to animals. Bioaccumulation factors may then operate in foodchains; and for some radionuclides in some ecosystems the result could be concentration of radioactivity highly adverse to particular species. Because of stratospheric transport of fallout, such "hot spots" could appear at distances far removed from the site of the originating nuclear explosion.

The indirect effects to which nuclear-weapons employment might expose ecosystems stem from the sheer magnitude of the energy-release from nuclear weapons. Batten[32] has estimated that, per megaton TNT-equivalent of yield, the cratering action of surface-burst nuclear weapons is likely to inject 1000–10,000 tons of dust into the atmosphere in the form of submicron aerosol. Such dust may provide condensation nuclei for cloud formation and may appreciably absorb radiation to or from the earth. Significant local weather changes might result; and, if the dust injection were sufficiently massive, its atmospheric influence might become manifest on a global scale. Nier and co-workers[33] estimate that the dust cloud from 10,000 megatons of nuclear explosion might reduce the average global surface temperature by some tenths of a degree Celsius over a 1–3-year period. They also estimate that about 10,000 tons of nitrogen oxides per megaton yield would be injected into the stratosphere from air-burst nuclear weapons where depletion, via photochemical reaction, of stratospheric ozone could result. On

---

[32] E. S. Batten, "The effects of nuclear war on the weather and climate", Rand Corporation memorandum No. RM–4989–TAB, August 1966.
[33] U.S. National Research Council, Committee to Study the Long-term Worldwide Effects of Multiple Nuclear-weapons Detonations (Chairman, A. O. C. Nier). *Long-term Worldwide Effects of Multiple Nuclear-weapons Detonations* (Washington, D.C.: National Academy of Sciences, 1975).

a sufficiently large scale, the effects could be a further reduction of average global surface temperature and an increase in the amount of ultra violet radiation in the biologically active wavelength range reaching the earth's surface from the sun. Both consequences could have extensive ecosystemic implications, especially the ultraviolet effects.[34] After a large-scale use of nuclear weapons, and during the 10-20-year period which would elapse before the stratospheric concentration of ozone returned to its former level, there could be worldwide effects on climate, crop-production, mutagenesis of pathogenic and other microorganisms, together with a marked increase in the incidence of fatally intense sunburn, skin cancer, etc.

---

[34]A convenient documented summary of current knowledge on the possible biological effects of increased ultraviolet-B irradiation due to reductions in atmospheric concentrations of ozone is contained in the *UNEP 1976 Annual Review* at pp. 11-15.

Chapter 3

ECOSYSTEMS

It is now necessary to find an appropriate way of defining the vulnerability of an ecosystem towards weapons, and then to marshal the available empirical information that bears upon the relative vulnerabilities of different ecosystems. For this process of definition, classification and description a brief preliminary review is made of some of the basic principles of systems ecology.

STRUCTURAL AND FUNCTIONAL COMPONENTS

An ecosystem can be looked at as a theoretical concept: an intellectual device for ordering and interpreting ecological data. Its use in describing those features of the natural world which it subsumes can be said to have imparted to it a sense of physical reality, so that the contents of a particular tract of land, for example, may be referred to as an ecosystem. In common with all such concepts—whether they be, say, the "molecule" of chemistry or the "state" of political science—an ecosystem is an abstract model of a part of the world.

The outlines of what ecologists mean when they use the term have been described in Chapter 1: essentially, a complex of interacting factors and organisms making up a definable part of the total environment. Just as no individual organism can be self-sufficient, neither can any one ecosystem. The largest and most nearly self-sufficient of all is the biosphere itself, whose survival depends upon a continual flow of energy between the sun and the thermal sink of space. Between more limited ecosystems, whether at the macro level of a continent or at the micro-level of, say, a pond or a meadow, there will be a continual flow not only of energy but also of materials and, to a greater or less extent, biota. The delimitation of an ecosystem is therefore to an extent artificial: a matter of convenience in which the choice of boundaries will largely depend upon the purposes for which the ecosystem is to be defined. Whether it is done geographically, operationally or by any other criterion, the factor of dependence upon externalities will persist. An important implication for the present study is that a man-made disturbance in any one ecosystem may propagate itself by natural means throughout contiguous ecosystems. The Earth is one huge *integrated* life-support system.

In whatever manner an ecosystem is delimited, it components may be grouped into categories that are common to all ecosystems. One such structural categorization, following Eugene Odum,[35] is this:

(a) Inorganic substances involved in material cycles: carbon, nitrogen, phosphorus, water, carbon dioxide, etc.
(b) Organic compounds: proteins, carbohydrates, lipids, humus, etc.
(c) Climate regime: temperature and other such physical factors.

These three categories comprise the abiotic components of the ecosystem; the other categories are the biotic components:

(d) Producers: organisms, principally green plants including algae, capable of fixing incident solar energy and using simple inorganic substances to build up complex organic substances that can serve as foodstuffs for other organisms.
(e) Consumers: organisms (mostly animals) which ingest other organisms or particulate organic matter.
(f) Decomposers: organisms (mostly bacteria and fungi) which decompose dead protoplasms, absorb some of the decomposition products and release inorganic substances that producer organisms can use as nutrients, together with organic compounds which may provide energy sources or which may be inhibitory or stimulatory to other biotic components of the ecosystem.

Any extent of space and time, however large or small, in which these six categories of structural component coexist and interact with one another to achieve some sort of functional stability may be regarded as an ecosystem.

An ecosystem may also be defined in functional terms. The functional

---

[35]E. P. Odum, *Fundamentals of Ecology* (Philadelphia: W. B. Saunders Company, 1971, 3rd edition), p. 8.

relationships between the structural components noted in the preceding paragraph may be grouped into categories of process that are common to all ecosystems. One such functional categorization, again following Odum, is this:

(a) Energy-transfer processes (energy circuits).
(b) Food chains.
(c) Biogeochemical nutrient cycles.
(d) Diversification, in space and time.
(e) Developmental and evolutionary processes.
(f) Control processes.

A stable ecosystem is one in which the first four of these processes have attained an approximate state of dynamic equilibrium with one another, the result in large ecosystems being that familiar "balance of nature" whereby the natural environment and its living community appear much the same year after year, inherently resisting change. Such homeostasis is the outcome of the control processes, interpretable in systems-analysis terms as feedback loops within the overall nexus of energy circuits, food chains, biogeochemical cycles and diversity patterns. A new ecosystem will come to display strong homeostasis only after a period of evolutionary adjustment comprising mutual development of its structural components. Each category of functional process comprises a subset of contributory processes. For (a), (b) and (c) these subsets include, as regards the biomass, photosynthesis, herbivory, predation, parasitism and symbiotic activities, and, as regards the geomass, evaporation, precipitation, erosion and deposition. For (d) and (e) the subsets include the growth and reproductive processes, the biological and physical agencies of mortality, immigration into the ecosystem and emigration from it, and the development of adaptive habits.

Before an ecosystem attains its climax state of maximum stability, it will have evolved through a succession of intermediate stages or "seres". This process of ecological succession results from modification of the physical environment by the community of populations inhabiting it. The modifications brought about by the growth of one type of population creates new niches in the environment favouring the establishment and growth of additional types of population, and so on until ultimately, at climax, full equilibrium between biotic and abiotic components is attained. Each sere will present an appearance of stability in that it will be characterized by a selection of species filling its niches that will alter only gradually as the years go by; it may thus be seen as a dynamic equilibrium that is slowly being displaced by pressure of evolutionary change. The classical studies here are those of plant ecology in the British Isles begun by Tansley in 1911[36], one of the basic successions which he identified may be summarized thus:

> Bare rock—wind erosion and windborne dust—holding enough moisture to sustain lichens—which die—humus and more retention of moisture—germination of moss spores—accumulation of soil—germination of seeds of herbs or spores of ferns—continuous plant cover—increased number of niches—increasingly varied plants—scrubland—trees—forests.

Succession is community-controlled within limits set by the physical environment. In cold or arid regions, for example, the forest stage may never be reached; the climax plant community may then be scrubland or some more primitive form of vegetation. Cultivated land may be seen as an ecosystem that man endeavours to shield from successional change by stripping and tending it so as to hold back species that would otherwise compete successfully with crop plants.

ADAPTABILITY AND VULNERABILITY OF ECOSYSTEMS

How far an ecosystem has developed towards its climax stage will profoundly affect the manner in which it adapts to perturbations by external factors, including assault by weapons. As ecological succession proceeds, the species populating an ecosystem becomes more diverse so that the ecosystem becomes better able to sustain fluctuations in species without substantially changing character. In other words, if the perturbation consists of catastrophe to one or a few populations, the more developed the ecosystem, the more likely are its control processes to restore its original state. Recovery may be relatively rapid. But if the perturbation is more severe — if there is greater depopulation of species or greater destruction of habit to which particular species are adapted—the homeostatic mechanisms may be over-ridden. Upheavals or other alterations in the environment of a community of populations may create conditions which only a few species are able to tolerate and therefore survive. The more sensitive or less adaptive species will then move away, perish or become unable to propagate themselves. The overall

---

[36] See especially A. G. Tansley, *The British Islands and their Vegetation* (Cambridge, 1939).

result will be a progressive increase in the numbers of representatives of tolerant, usually more primitive, species. The entire ecosystem may then, in effect, be forced back into an earlier sere. In such an eventuality, recovery may be very slow, proceeding no faster than the evolutionary process of successional change.

Precisely what sort or what degree of damage to the structural components of an ecosystem would constitute a perturbation of the latter severity will clearly vary from ecosystem to ecosystem. Not only will it be a matter of stage of ecological succession; the prevailing physical environment, itself setting limits to succession, will also be critical. Thus temperate ecosystems, which in general contain a larger number of ecological niches available for population, will be inherently more adaptive to perturbations than arctic ecosystems, which contain few niches.

The adaptability of ecosystems towards perturbation by weapons may be taken as an indication of their vulnerability. Ecosystems may then be differentiated as regards vulnerability according to the manner of their adaptation: the rate at which they return to an equilibrium condition, and the extent to which the new equilibrium differs from the old. Such a differentiation requires some way of measuring, or at least ranking, rates and directions of adaptation. This can be done by focussing on changes in particular components of ecosystem structure or on particular measures of ecosystem function.

For this there are many alternative possibilities, but for present purposes it would seem that man, considered as one species of organism making up the consumer community of an ecosystem, offers a suitable reference point. Ecosystem vulnerability can then be assessed in terms of the extent to which the rate and direction of adaptation by the ecosystem restores damage done to the human habitat by weapons assault. A further reference point is needed in order to specify what constitutes damage, for degradation of habitat may affect the human population in ways that range from direct and immediate threats to survival to the creation of new constraints on economic and social development that may initially be imperceptible. Since the present study is motivated by concern about one particular form of possible damage, that of desertification, it is this that will be taken as the second point of reference. Special significance must be attached to the time factor, for though an ecosystem may ultimately recover, in one fashion or another, it will be the rate at which it does so in terms of human life-spans that will be immediately critical for the human population.

CLASSIFICATION OF ECOSYSTEMS

This manner of specifying vulnerability also indicates the manner in which ecosystems should be differentiated from one another for descriptive purposes in the present study. The ecosystems delimited must be large enough to subsume communities of people within their biotic components and to admit considerations of, for example, climate and habitat at a level of aggregation commensurate with current knowledge about desertification processes. This points to a classification of ecosystems by means of a climatic-geomorphic differentiation of habitats. One such classification is a division of the global ecosystem into six categories of subsidiary ecosystem: temperate, tropical, arctic, arid, insular and oceanic. Table 3 shows the approximate total area of the earth's surface that may be taken as falling within each of the six categories, together with the approximate total human population of each category and the numbers of nations which those populations constitute.

A major difficulty has to be faced in reconciling the generality of this ecosystem classification with the specificity that characterizes the current state of knowledge in ecology. Ecologists have tended to concentrate on much smaller ecosystems. This means that while a considerable amount is known about the functioning of some parts of each of the six categories of ecosystem, other parts have received rather little attention, and the whole still less. There is thus no solid body of scientific knowledge about the vulnerability of ecosystems conceived at the macro level from which to make reliable assessments. Much of what follows can therefore be taken only as generalization from impressions.

Even so, it is reasonably clear that the magnitude of the possible human consequences from weapons-assault varies over the six categories of ecosystem. This means that a ranking of potential vulnerability can be made. As seen on page 32, the measure of vulnerability is taken as the severity of ecosystemically-mediated harm likely to be suffered by human populations directly related to the ecosystems. In this special sense, therefore, oceanic ecosystems, which have no human inhabitants, may be placed at the bottom of the scale of vulnerability. At the opposite end are the arctic, arid and tropical ecosystems which, for reasons primarily of species-poverty in the first two and soil-weakness in the last, have a degree of brittleness which could greatly reduce the capacity of their human inhabitants for withstanding environmental damage due to weapons. Some, though not all, insular ecosystems may present a

TABLE 3. SUBSYSTEMS OF THE GLOBAL ECOSYSTEM

Category of Subsidiary Ecosystem	Total surface Area (millions of square kilometres)	Total human component	
		Population (millions of people)	Communities (number of nations and other political entities[a])
Geomorphic classification			
Oceanic	361.3	0	0
Continental	141.5	3,631	125
Insular	7.2	434	34
Climatic classification (Land)			
Temperate[b]	56.2	2,360	56
Tropical[b]	42.2	1,586	80
Arid[c]	26.7	102	22
Artic	23.7	18	1

[a]Wholly or primarily within each category.
[b]Excluding regions classifiable on climatic data as "arid" or "hyper-arid".
[c]Excluding regions classifiable on climatic data as "semi-arid".

Source: Adapted from A. H. Westing, "The military impact on the human environment", in Stockholm International Peace Research Institute, *SIPRI Yearbook 1978* (Stockholm: Almqvist & Wicksell, 1978), pp. 43–68.

comparable degree of vulnerability, their relative isolation from other terrestrial ecosystems giving rise to certain peculiar fragilities. In temperate ecosystems the extensive industrial and technological base which sustains many, though not all, of the human inhabitants may afford a degree of resilience lacking in other ecosystems. These are all propositions which will now be discussed in greater detail.

ARID ECOSYSTEMS

Table 3 shows that arid ecosystems constitute about 18 percent of the world's land surface. A similar percentage is classifiable as "semi-arid". It is estimated that some 14 percent of the world's population lives in these arid and semi-arid regions.[37] Their locations are most clearly shown on the World Map of Desertification[38] produced for the UN Conference on Desertification by FAO and UNESCO in co-operation with WMO and UNEP. This map depicts five main desert belts ("deserts" being defined in the Explanatory Note accompanying the map as "regions where vegetation is scarce or absent because of deficient rainfall or edaphic aridity") around which the principal inhabited drylands occur: (1) the Sonoran desert of northwestern Mexico and its continuation in the desert basins of the southwestern United States: (2) the South American Pacific coastal desert, a thin coastal strip running west of the Andes from southern Ecuador to central Chile, whence dry climates extend eastwards into Patagonia; (3) a vast belt running from the Atlantic to China and including the Sahara, the Arabian desert, the deserts of Iran and the USSR, the Rajasthan desert of Pakistan and India, the Takla-Makan desert in China, and the Gobi desert in China and Mongolia; (4) the Kalahari and its surrounding arid lands in southern Africa; and (5) most of Australia. Outside these principal desert regions are isolated areas of arid lands in many parts of the world, such as southwestern Madagascar, part of northeastern Brazil and the Guajira Peninsula in Colombia.

The dominant characteristic of arid ecosystems is the dearth of water. This is a consequence of high rate of evaporation, scant rainfall and inadequate soil structure. The high evaporation is due to a combination of factors, the incidences and strengths of which vary from one region to another: high diurnal temperature, low air humidity, strong winds and cloudless skies. Since water is a physiological necessity for all living organisms, arid ecosystems have a low biomass; and their sparse flora and fauna are highly specialized for arid regions. In their adaptation to extreme conditions, the producer organisms are of a kind such that a relatively large amount of the net production of arid ecosystems goes into storage or reproductive organs. This appears to favour permeant consumer organisms, such as herbivorous rodents (though it has to

---

[37]M. Kassas, "Arid and semi-arid lands: an overview", in UNEP, *Overviews in the Priority Subject Area: Land, Water and Desertification* (Nairobi, February 1975). '
[38]A/CONF.74/2.

be said that the studies that have been made on ecosystemic function in arid regions are few compared with purely descriptive studies). Such fauna, though limited, include species that can live indefinitely on, for example, dry seeds: this may well play an important part in nutrient cycling. If so, the fact that consumers may be relatively more important in this respect than bacterial or fungal decomposers in other ecosystems may well introduce a significant locus of vulnerability towards mass-destruction weapons. It may be noted that one of the more commonly observed factors in the desertification of semi-arid regions is overgrazing and over-browsing by another category of permeant consumer, namely the livestock animals of nomadic peoples and sedentary pastoralists.

In irrigated arid lands, the nature of the soil becomes a primary factor in ecosystem function and stability. Where texture and nutrient content of the soil are favourable, the large amount of sunlight may lead to high productivity. By the same token, damage to soil structure may swiftly degrade semi-arid ecosystems by decoupling hydrological cycles which, in biologically sparse regions, may have little more than the physical and chemical composition of the soil to sustain them. Such degradation could prove to be accelerating in character, for it is well observed that an increase in arid-ecosystem productivity due to an increase in moisture takes the form of a rapid succession of desert species of plant by other species adapted to greater moisture: the enforcement of a low-moisture regime by soil damage could thus be expected to kill off most of the newly established primary producer population. Herein may lie a second important form of vulnerability.

Soil stability in any ecosystem is enhanced by vegetational cover. Since in arid regions the cover will in any case be sparse most of the time, and since it also represents the primary producer component of arid ecosystems, such sensitivities as it may display towards degradative influences may constitute a synergetic form of vulnerability. It may also mean that a semi-arid or sub-humid ecosystem which has been stripped of its vegetation may take a long time to regain, if it does so at all, a state of dynamic equilibrium comparable with the initial state. In which case, when considered within the time-frame of human generations, the desertification may be permanent rather than temporary.

TROPICAL ECOSYSTEMS

Table 3 shows that about 40 percent of the population of the world lives within tropical ecosystems, these constituting rather less than 30 percent of the world's land surface. Many of the 80 nations into which this population is organized are exceedingly poor. The three major tropical regions are in central Africa, southeastern Asia and northern South America. Some three-quarters of the lands are forested, most of the remainder being grassland characterized as tropical savannah. Rather less than half of the forested lands constitute closed tropical rain-forest; the remainder, in which the moisture conditions are intermediate between savannah and rain forest, comprise open forests of various types, such as the scrub and thorn forests of the African bush and the deciduous forests of the Asian monsoon region.

The complex interaction of warmth, rain and evaporation in tropical rain-forest enables the leaves to perform an extraordinarily efficient solar energy conversion, so much so that the gross primary productivity of a tropical rain-forest may exceed that of the most advanced agricultural methods in practice today. This phenomenon sustains a pattern of nutrient cycling that sets tropical ecosystems apart from other categories of ecosystem. The main characteristic is that a much larger proportion of the organic matter and the available nutrients in a tropical ecosystem, and above all in a tropical rain-forest, is held in the biomass rather than the soil, and recycled within the organic structure of the ecosystem. It seems that tropical rain-forests have developed special mechanisms whereby mineral nutrients in the free, inorganic state are denied full access to the soil, where they would rapidly be leached away by the heavy rainfall. Went and Stark[39] have postulated, for example, that the extremely abundant mycorrhizae (root fungi) of the surface litter and the thin humus of the forest floor are capable of digesting dead organic litter and then passing minerals and food substances through their hyphae directly to living root cells. If such recycle pathways via symbiotic micro-organism links are indeed of major functional importance in tropical eco-systems—and the available evidence suggests that they are—they may also represent a major locus of vulnerability.

Tropical savannah is characterized by a labile equilibrium between grassy and woody species, which are antagonistic plant types. Walter[40] has demon-

[39] F. W. Went and N. Stark, "Mycorrhizae", *Bio-Science* 18: 1035–1039 (1968).
[40] H. Walter, *Vegetation of the Earth* (translated by J. Wieser from the second German edition; London: English Universities Press, 1973), pp. 67–71.

strated that the reason why an equilibrium between them can be reached at all lies in the tropical coincidence of summer rains and deep, loamy sand; and that the lability of the equilibrium lies in the different root systems of the two categories of producer organism and their water economy. Thus, only when the annual rainfall exceeds about 300 mm do the intensive root systems of the grasses permit adequate passage of moisture to the extensive root systems of woody plants thereby enabling the latter to survive the dry season. Heavier rainfall may create sufficiently favourable conditions for tree as well as shrub woody species; and once the state is reached of tree crowns merging to form a canopy, thereby overshadowing grassy photosynthesis, the competitive relationship is reversed, and woody plants come to dominate the ecosystem. It is the grasses, however, that are the more important producer as far as the cattle-reliant human population of the ecosystem is concerned. Through their labile equilibrium with the woody species, the grasses thus present a double locus of vulnerability, for if they are removed (as by fire or overgrazing), woody plants will develop luxuriantly, their seedlings no longer exposed to grass-roots competition: a thorny scrubland may then rapidly become established, useless for grazing purposes. Brush encroachment of this kind may be a prelude to one particular well-observed form of man-induced desertification.

Other forms of vulnerability may be envisaged with regard to those of the human populations of tropical ecosystems that rely on agriculture rather than animal husbandry. Ecosystemically-mediated, as opposed to direct, damage to crop cultivation may result from soil deterioration due to any one of a number of disturbances, whether pedological, hydrological or vegetational. While the same is true for any ecosystem, those of the tropics have certain peculiar fragilities. For example, in savannah regions where the annual rainfall is sufficiently high to support agriculture and where there is a distinct dry period, fire may seriously menace cultivation, not only directly, but also indirectly. Large-scale, intense fires may so scorch the soil as to degrade aeration, water-storage capacity, nutrient retention and sub-soil biomass, thereby decoupling transfer cycles across wide expanses of the ecosystem in both space and time. Previously fertile areas may thus be rendered barren, and the attendant possibility of erosion of the exposed topsoil under the influence of wind or rain may greatly, even permanently, delay restoration of an equilibrium that favours agriculture. Successional change during the intervening period may take a variety of forms, some of which could be actively inimical to the human population. A common characteristic of burnt-out areas is, for example, a rapid colonization by insect species. Dependent upon the nature of the colonizing species, this may not only inhibit resumed attempts at agriculture, but may also lead to the establishment of new foci of human or animal disease.

ARCTIC ECOSYSTEMS

Arctic ecosystems account for about 16 percent of the total land area of the earth. The 24 million km^2 of lands which this comprises are covered with snow and ice either permanently or seasonally, mostly because of latitude but also, in the case of about 2 million km^2, because of altitude. Their human population, part of which leads a traditional nomadic life, represents only about 0.4 percent of the world population. Most of it exists within the tundral zones that cover about 6 million km^2. Human survival within the frigid deserts that comprise the greater part of the arctic and antarctic regions is exceedingly precarious.

Cold is the determining factor in arctic ecosystems. Relatively few kinds of organism can adapt themselves to low temperatures. The consequently low species-diversity of arctic ecosystems imparts a particular fragility. Primary productivity is low. Retardation by the cold of microbial decomposition and other factors preclude rapid cycling of nutrients. The foodchains are short and the choice of food for consumer organisms, including man, is very limited, so that violent fluctuations in the size of some populations may be frequent. Adaptation to disturbance is slow.

The soil of tundral ecosystems is permanently frozen except for an uppermost layer that thaws each summer. It sustains, and is insulated by, a mat of low-growing vegetation comprising lichens, grasses, sedges and dwarf woody plants. The lichens, with their strong capacity for metabolizing inorganic minerals, including airborne minerals that become deposited upon them, assume a particular importance in nutrient cycling under such circumstances. Herein lies one of the more familiar examples of ecosystem vulnerability to weapons effects, for this metabolic capacity has proved to function as well with radioactive fallout from atmospheric nuclear-weapons testing as it does with natural windborne inorganic substances. The result has been that radionuclides, including those that are stongly retained by animal tissue (strontium-90 and caesium-137), are swiftly incorporated into the short foodchains upon which the human population relies. The lichen-reindeer/caribouman foodchain constitutes a particularly powerful bioaccumulator of radioactivity.

INSULAR ECOSYSTEMS

Fosberg[41] and Westing[42] have drawn attention to the characteristic vulnerability of oceanic-island ecosystems. The smaller oceanic islands have in common a physical habitat that is severely circumscribed, so that the total number of species of living organisms on each island is limited in comparison with a mainland of equal size. Some of these species may be unique to the islands which they inhabit; and since there will be less pressure of interspecific competition than on a mainland, bizarre and ill-adapted forms have a higher likelihood of survival and make a greater contribution to the functioning of the overall ecosystem. However, precisely because of the ill adaption of such species they may prove peculiarly vulnerable to abnormal stresses, thereby rendering the entire ecosystem fragile to external disturbance. Such fragility will be exacerbated by the general lack of robustness characteristic of any ecosystem which is low in species diversity.

TEMPERATE ECOSYSTEMS

Temperate ecosystems now subsume most of the technically advanced nations. They comprise regions of the world where the natural environment has been the most extensively and drastically modified to suit human needs. This has mainly taken the form of increasingly intensive industrialized, agricultural and silvicultural enterprise, and high-density urbanization. The potential for ecosystemically-mediated damage to the human population therefore takes on a different character from that in other ecosystems. There is a variety of man-made features which reduce fragility, above all the deliberate and massive coupling of key energy and material circuits into the natural resources of external ecosystems. This may serve to cushion temperate ecosystems against disturbance, thereby reducing vulnerability. But, at the same time, the external couplings could themselves constitute vulnerability under some circumstances; and, as the survival of the human populations of temperate

---

[41] F. R. Frosberg, "Man's effects on island ecosystems", in M. T. Farvar and J. P. Milton (eds.), *The Careless Technology: Ecology and International Development* (Garden City, New York: The Natural History Press, 1972), pp. 869–880.
[42] A. H. Westing, "The military impact on the human environment", in Stockholm International Peace Research Institute, *SIPRI Yearbook 1978* (Stockholm: Almqvist and Wicksell, 1978), pp. 43–68 at 61–63.

ecosystems becomes dependent increasingly upon technology and decreasingly upon "natural" forms of social organization, disturbances that could not be cushioned might become transformed into human catastrophe on a scale far surpassing that in other ecosystems.

Chapter 4

POSSIBLE IMPACTS OF WEAPONS ON ECOSYSTEMS

In order to assess the possible impacts of weapons on ecosystems, the weapons-effects characteristics set out in Chapter 2 may be related to the ecosystem-vulnerability characteristics set out in Chapter 3. A limited amount of empirical evidence is available to guide such a theoretical analysis, and it is useful to begin by reviewing it briefly.

EMPIRICAL DATA

It must be stressed that the empirical evidence does not go very far. A sizeable body of data exists concerning the impact of war on individual features of the natural environment:[43] there are recorded observations on changes in, for example, the local topography of battle zones and in the populations of certain of the more conspicuous species of living organisms. Such descriptive information is not at all the same thing, however, as knowledge of the ecological impact of war, still less of the impact of weapons on ecosystems. Ecological surveys and analyses of the kind required for such systematic understanding have rarely been attempted; and on the few occasions when they have, a shortage of resources and inadequacies in baseline data and methodology have severely circumscribed the insight gained. It seems, moreover, that they relate to the Viet-Nam conflict and are therefore confined to one particular set of geomorphic and climatic circumstances and to a limited range of weapons.

[43] A bibliography has been compiled: No. 40 in the UNESCO *Reports and Papers in the Social Sciences* series, "The danger to man and his environment inherent in modern armaments and techniques of warfare: an introduction and descriptive annotated bibliography" (Paris: UNESCO, in press).

The only other body of ecological observation which significantly supplements the available empirical evidence relates to weapons proving grounds; but this, too, is severely limited in scope and much of it remains unpublished.

With regard to biological weapons, empirical data bearing upon ecosystemic effects derive from ecological surveys of past test areas. But little information has been published about such testing and practically nothing reporting observations made during or after it. Arctic, arid, insular and oceanic ecosystems have all provided test sites. In at least one of them, test micro-organisms (anthrax spores) remain viable in the soil 36 years after cessation of testing.

As to antipersonnel chemical weapons, it seems that only one attempt has been made to track ecological sequelae of their massive employment in the temperate ecosystem of Europe during 1915-18, and that nothing significant was found. Test-area observations remain, for the most part, unpublished, though there are exceptions.[44] Arctic, arid, tropical, temperate and insular ecosystems have all provided test sites.

Considerably more empirical data are available on the ecological effects of antiplant chemicals. Many of them are derived, however, from civilian agricultural, silvicultural and range-management experience with herbicides; although this experience has been extensive, its relevance is limited by the military requirements for rapidity of effect and effectiveness across a broad spectrum of plant species, these necessitating application-densities far exceeding that of civil practice. There have been at least four wars since World War II in which herbicides have been employed offensively, but only for the herbicide operations of the Viet-Nam conflict, which were by far the largest, have attempts been made to assess ecological impact. The field-work on which was based the fuller of the assessments that have thus far been published was all conducted while the war was in progress, and the consequent constraints both on access and on time-frame permit only tentative and incomplete conclusions. It may be a matter of decades before the full ecological impact has become manifest and can, within the limitations of available baseline data, be assessed.

Six separate investigations of the ecological effects of military herbicide operations were conducted by US scientists in Indochina during 1968-1973,

---

[44]See, for example, F. P. Ward, "A summary of ecological investigations at Edgewood Arsenal, Maryland: fiscal year 1970", U.S. Army, Edgewood Arsenal, Special Publication No. EASP 100-101, June 1971.

and in 1977 a further study was undertaken as part of the work of the Special Mission designated by the UN Secretary-General in compliance with General Assembly resolution 32/3 of 14 October 1977 for the purpose of reviewing the progress of international assistance for the reconstruction of Viet-Nam. Literature references to the reports of this work are listed in the Annex. A review[45] of the findings of the first six investigations has been published by Professor Arthur Westing, the forester, who was involved in three of them; this review contains the following commentary:

> "Among the ecological lessons to be learned from the chemical antiplant warfare in Indochina are: (a) that the vegetation can be utterly destroyed with relative ease over extensive areas; (b) that this in turn has a devastating impact on the animal life depending upon this vegetation for food or shelter; (c) that the ecosystem is, through such attack, subject to nutrient dumping (that is, to the rapid and major loss of soluble nutrients); and (d) that the ecological debilitation from such attack is likely to be of long duration. To this list can be added a number of social lessons to be learned as well, aming them: (a) that natural, agricultural and industrial-crop plant communities are all similarly vulnerable, and (b) that the local civil population can suffer extensively from such action in a variety of direct and indirect ways."

Some of the basis for these remarks is described later in this chapter. The report by the UN Special Mission follows Westing in estimating that the total area of south Viet-Nam damaged by herbicides amounted to somewhat more than 17,000 km^2, including 1510 km^2 of completely destroyed mangrove forest.

With respect to the environmental effects of nuclear weapons, empirical data are to be found in the surveys that were made in Japan, after the atomic-bobming of Hiroshima and Nagasaki. Much of this information is reported in the review by Glasstone[46] of the effects of nuclear weapons. Richer as sources of ecological data are the surveys that have been made at above-ground nuclear-weapons test sites, of which the better known ones have been located within arid or insular ecosystems. Biological and ecological studies have been reported

---

[45]*Supra*, note 24, at pp. 24–25.
[46]S. Glasstone (ed.), *Effects of Nuclear Weapons* (Washington, D.C.: U.S. Atomic Energy Commission, revised edition, 1964).

from the test sites in the Mohave Desert, Nevada, U.S.A., and in the tropical Pacific area; a review has recently been published by Westing.[47]

As to the Mohave Desert test area, in which there had been at least 89 above-ground test explosions of yields in the 10–70 kiloton range, it is recorded that for none of the explosions did the area over which initial vegetational damage could be detected exceed 3255 hectares; within that area, the zone of severe vegetational damage was between 400 and 1375 hectares for each explosion, this having a central zone in which the destruction of life was essentially complete ranging in area from 73 to 204 hectares. Within 3–4 years, pioneer plant species had invaded the central zones, with slow ecological recovery under way in the adjacent zones. The pattern of ecological succession appeared to be similar in kind and rate to that following any severe disturbance of desert-region habitat: a process spanning many decades.

As to the Pacific-island test sites, vegetational recovery was also observed, apparently following the normal and, in this case, relatively rapid, successional course. However, local extinction of at least one animal species was found. Certain radionuclides, including caesium-137 and strontium-90, appeared two years after testing to have permanently established themselves in at least one biogeochemical circuit of the nutrient cycle. Recent studies have shown abnormally high amounts of these two isotopes, as well as plutonium, in the bodies of Bikini islanders.

With regard to the effects of conventional weapons, the reported environmental data are, as noted on page 42, extensive but as yet too fragmentary and unsystematically collected to be of value here. The picture may possibly be clarifed by a comprehensive review of the subject that is now in the process of publication.[48] As with chemical herbicides, there are parallel experiences from which pertinent findings may be drawn in pest-control, land-management and agricultural activities that have made extensive use of fire or explosives. Notable here are the attempts at controlling Weaver-bird (*Quelea quelea*) colonies in West Africa, in particular the heavy use of napalm and defoliants during the OCLALAV control programme of the 1960's in Senegal.

---

[47] A. H. Westing (for the Stockholm International Peace Research Institute), *Weapons of Mass Destruction and the Environment* (London: Taylor & Francis, 1977), pp. 20–21.
[48] A. H. Westing, for the Stockholm International Peace Research Institute, *Warfare in a Fragile World* (London: Taylor & Francis, in press).

THEORETICAL FRAMEWORK

The weapons-effects characterizing the weapons treated in Chapter 2 have been described in terms of the damage-agents of the weapons: high kinetic-energy fragments, blast, flame, heat flux, toxic and radiotoxic substances, pathogenic microbes and ionizing radiation. Though their actions upon biological systems differ greatly from one another when considered at the level of individual organisms, they have features in common when considered at the higher level of the community. It is possible, therefore, to assess potential impacts of weapons on ecosystems in terms of a relatively small number of stress factors, each one representing the aggregate effect of one or more of the relatively large number of different damage agents.

Several such aggregations are possible. In accordance with the reference points adopted in this study for assessing ecosystem vulnerability (see page 32), the most useful aggregation is that which bears most directly on the mechanisms of desertification. All terrestrial ecosystems are potentially reducible to deserts or to other such states of low biological productivity, incapable of supporting sufficient plant life to cover more than small fractions of their surfaces. This may be because there is no soil, no water, too much water, or too much salt or other phytotoxicants; because the prevailing temperature is too high or too low; or because the soil structure is inadequate for holding water or nutrients. The conceivable mechanisms of desertification and analogous processes due to warfare follow from this. An indirect example would be oversalination due to inadequate irrigation following damage done to the means of irrigation. A direct example would be destruction of soil structure. The latter might be brought about by repeated or drastic mechanical disturbances, including frequent trampling or passage of heavy vehicles; by extremes of temperature, especially alternating extremes; or by the killing of the microbiota of the topsoil. The capacity of weapons for bringing about desertification would then depend in large measure on their propensity for damaging the soil or killing its biota, effects which would be exacerbated by destruction of the plant cover protecting topsoils from extremes of temperature and from desiccation.

It would seem, then, that the most appropriate aggregation of weapons-effect damage agents is into the three composite stress factors of soil damage, destruction of plant cover, and biocide. All of the weapons under consideration may stress ecosystems in one or more of these three ways. Some will do so more than others. Thus conceived, the stress factors are not independent of one anoher. Some forms of biocide may, for example, result in a diminution of plant cover which may in turn bring about damage to the soil. Such dependencies may be captured by differentiating direct and indirect stress and considering how the latter may affect resilience towards the former.

As an analytical tool, this notion of stress permits generalizations to be made about impacts of different weapons on each of the different categories of structural component which together, as described on page 29 constitute an ecosystem. A similar transformation may be applied to the ecosystem-sensitivity characteristics set out in Chapter 3 for these can be conceived as weaknesses in certain of the functional links which together, as described on page 30, define an ecosystem. Conclusions about possible impacts of weapons on the different categories of ecosystem under consideration may then be drawn in terms of the degree to which the stresses acting on the structural components are likely to aggravate the functional weaknesses, thereby threatening ecosystemically mediated harm to the human populations of the ecosystems.

STRESS THROUGH SOIL DAMAGE

The soil is the vital link between the biotic and abiotic components of an ecosystem. It acts as a reservoir for water and the other inorganic substances cycling through the ecosystem. It provides habitat for many of the different populations of decomposer organisms that control the rate and capacity of nutrient cycling. It affords the physical structure in which are rooted most, if not all, of the different populations of primary producers. Soil damage may thus stress an ecosystem in a wide variety of ways. Two broad types of soil damage may be envisaged as a consequence of weapons employment: physical displacement of soil; and alterations in soil structure and composition.

Soil displacement may be a direct consequence of the cratering action of the blast from high-explosive or nuclear weapons. Or it may be an indirect consequence of the destruction of vegetation whose loss subsequently causes increased exposure to the erosive influences of wind and water. The uppermost layers of soil are at once the most vulnerable and the most immediately important for the ecosystem as a whole. They comprise a topmost layer of vegetational litter and other detritus; an intermediate layer of organic matter largely comprising the humus resulting from microbial decomposition of the surface litter; and lower layers of soil having a high porosity as a result of root growth and the burrowing actions of small subterranean creatures. It is this porosity and the crumb-sizing associated with it which enables the soil to retain, according to its chemical and physical nature, supplies of inorganic nutrients in readily assimilable form; in concert with the large water-holding capacity of the litter and humus, it provides the mechanism for infiltration of rainfall into the soil, rather than erosive surface flow. The thickness, content and structure of all these layers will vary greatly from one ecosystem to another. In general, the thinner the layers, the more vulnerable will be the entire ecosystem to weapons that cause soil displacement.

At any given moment, the topsoil present in a particular area represents an equilibrium between natural processes of soil erosion and soil genesis. Therefore, if weapon-effects that cause soil displacement also reduce the natural rate of soil genesis, the result may be an accelerating erosion of topsoil, which is the classic mechanism of desertification. Destruction of vegetation over a wide circumadjacent area will reduce soil-genesis rates. Eroding topsoil becomes wind- or water-borne particulate material which, in large quantities, may menace downwind or downstream areas. The mechanisms of, for example, dust-storms or silting may thus transmit the damaging consequences of weapons from one region to another.

The ecosystemic consequences of soil displacement may not be limited to those of topsoil damage. In some terrains, deep cratering may disturb the water-table and local drainage patterns. If a hardpan is breached, the perched water-table may be lost for many years.

The other major form of soil damage to be considered here comprises a variety of adverse changes in soil structure and composition. Thus, the chemical exchanges which take place in the soil as part of, for example, nutrient cycling occur by way of finely poised chemical and physicochemical equilibria such that small alterations in, say, the sorptive characteristics or the acidity of the soil may tilt the equilibria and block certain of the exchanges. Damage of this type is perhaps more likely to be an indirect than a direct consequence of weapons use: a consequence of, for example, accelerated weathering or other such chemical and physicochemical changes brought about by increased exposure due to cratering or loss of plant cover; or gross damage to constructions such as dykes which then admit increased salinity.

Some soil types may be especially sensitive. The latosols, which occur most frequently in the tropical belt within 30 degrees of the equator, are an example. For reasons which remain to be fully elucidated, latosols have a peculiar propensity for assuming a rock-like hardness upon prolonged exposure to the elements. The resultant laterite is a mixture of aluminium and iron oxides and very little else, incapable of supporting either natural growth or agriculture. Such induration is a mechanism of desertification that has confounded ill-conceived attempts at agriculture in a number of forested tropical regions.

Direct changes in soil structure and composition might possibly be brought about by chemical weapons, though it seems improbable that any of the chemical-warfare agents that are prominent today could cause such damage at their likely application densities. It must be noted, however, that the effects of these agents on soil microflora and microfauna do not appear to have been closely investigated. For the present it would seem that thermal stress, rather than chemical stress, is the more serious direct danger. Thus, the heat-flux resulting from the direct action of burning incendiary agents or from thermal irradiation by nuclear explosions, or from spreading fires initiated by either agency, may be of sufficient magnitude to bring about chemical and physical changes in the soil that reduce sorptive capacity and cause collapse of pore structure. Organic particulates may be oxidized and thereby lost from the ecosystem. Habitats for soil fauna my be eliminated. Soil scorching may therefore constitute a major stress on ecosystems.

STRESS THROUGH PLANT-COVER DESTRUCTION

Vegetation is the primary source of utilizable energy for an ecosystem. It is the first link in all the food-chains. It is the means whereby an ecosystem extracts inorganic minerals vital for its biotic components from the geomass. It provides not only food (directly or indirectly) for the consumer population, but also shelter. It stabilizes and moulds the soil, and transforms the prevailing climate regime into a microclimate favouring not only itself but also other biotic components of the ecosystem. The character of an ecosystem is thus dominated by its vegetation, so that damage to the latter will profoundly affect the former.

Destruction of plant cover might be brought about directly by the physical agencies of blast, fragment kinetic-energy, or fire. Or it might be caused indirectly either by soil displacement or by biocidal action upon the component flora. All the weapons under consideration may therefore expose ecosystems to the stress of plant-cover destruction.

The observable consequences of that stress may be diverse, much depending upon the scale of the vegetational destruction. If it is sufficiently large, chains or events will be set in process culminating in a variety of structural changes throughout the ecosystem. Their full range cannot be described here; but the following examples illustrate them.

The impact of plant-cover destruction upon the integrity and structure of the local soil has been suggested on pages 47 and 48; increased exposure of the soil to natural erosive forces and to leaching; retardation of soil genesis; and, in latosols, soil induration. Such might be the consequences of complete removal of plant cover. Less drastic reductions in plant cover might also profoundly affect the soil. For example, transpiration of water in an ecosystem would be reduced approximately in proportion to the surface area of damaged foliage. Though some increase in evaporation from the soil surface might occur, the overall evapotranspirative loss of moisture from the soil would in most cases be reduced. One possible consequence would be a generally moister soil and a rise in the local water-table. This could adversely affect the decomposer habitat and the productivity of plant species adapted to the pre-existing lower humidity. Where rainfall is substantial, the increased occupation of the soil's water-storage capacity would also have the effect—well observed by, for example, foresters in temperate regions—of forcing subsequent rainfall out of the affected area via increased surface or subsurface flow. The flood potential of the region would thereby increase and, with it, the possibilities of erosion and massive leaching of nutrients from the soil.

The foregoing effects may be seen as one possible set of consequences of alterations in microclimate due to plant-cover depletion. Such alterations, in addition to decreased humidity, would also include reduced surface carbon-dioxide concentrations (leading to diminished local photosynthesis) together with increased surface insolation and wind-velocity. The diurnal fluctuations of temperature gradient across the air layer immediately above the soil would become substantially more intense, and perhaps actively inimical to emergent plant seedlings and small permeant organisms.

Diminished vegetation may alter the reflectivity of the terrain towards solar radiation. Albedo change may provoke microclimate change.

In seeking to gauge the overall impact of stress by plant-cover destruction on ecosystem structure, a major consideration must be the rate and form of likely vegetational recovery. In broad terms, recovery may be seen as a two-phase process. The first phase comprises colonization of the denuded area and the establishment of a pioneer community. The second phase commences with the start of those successional changes noted on page 30 in which the pioneer community gives way to seres of increasing species diversity until the final climax community is reached, one which perpetuates itself through homeostatic mechanisms no longer perturbed by the forces of successional change. The closer the original community to a climax state, the more resilient may it prove, by virtue of its diversity, in the face of any of the stress-factors under consideration. But if the stress is great enough to overload the homeostatic mechanisms—as plant-cover destruction might be if it occurred on a massive scale—the net effect would, as noted on page 31, be reversion to an earlier and more primitive sere. The more highly evolved the ecosystem, the longer will it take to recover from such a reversion. In any event, the rate of recovery would depend upon the robustness of the ecosystem concerned, and the process could not commence without the initial phase of recolonization.

Unless the plant-cover destruction were slight, the new vegetation that came to establish itself in the area would generally be different from the original vegetation, having regard to the alterations that would have taken place in microclimate, nutrient budget, soil moisture-holding capacity, etc. It would establish itself through invasion by *Fusaria* (slime fungi), microscopic algae and bacteria, followed in succession by lichens, mosses and vascular plants. The new plant cover would consist of tenacious and undemanding grasses and similar species. Its overall character would be one of much lower biomass and nutrient-retention capacity. If the stress-factors acting upon the ecosystem had rendered it peculiarly inhospitable, revegetation might be delayed indefinitely, a consequence manifest as desertification. How likely such an eventuality might be would of course depend upon circumstances. The early colonisers would be windborne organisms; the extensiveness of the devastated area would control the appearance of the later ones. As noted on page 45, recovery is observable in certain arid and tropical-island ecosystems subjected to the massive insult of nuclear explosions. But by way of contrast, there is now the existence of the huge expanses of muddy wasteland that were once, prior to military herbicide operations, lush mangrove forests around the southern coastline of Viet-Nam. Precisely why little if any recolonization of these mangrove sites has yet occurred, at least in the form of vegetation, remains obscure: Testimony at once to the undeveloped state of ecological knowledge and to the gravity of possible ecosystemic consequences of plant-cover destruction.

STRESS THROUGH BIOCIDE

All weapons are capable of killing: of exposing living organisms to various kinds of biocidal action. Some are not in fact designed to do so, despite being capable of it. What the military call "antimaterial" weapons, for example, are designed to destroy inanimate structures or equipment, but the stresses which the weapons must exert to do so are invariably of such a kind and magnitude as to be biocidal for some organisms. Different species will differ in their robustness or fragility towards a given weapon-effect; and some types of weapon-effect will kill a wider variety of species than others. Hence, across the broad spectrum of living organisms, weapons will vary in the bandwidths of their biocidal activity. The precise extent of that bandwidth will be a factor of major significance in the degree to which the biocidal action of a weapon will stress ecosystems. Existing knowledge in this area, is, however, very incomplete, as the paucity of data on the matter given in Chapter 3 indicates.

In addition to the dominant primary producers, other organisms among the biotic components of an ecosystem may be especially important to the ecosystem as a whole. It may happen by chance, and could happen by design, that the more vital organisms fall within the biocidal bandwidth of a weapon. The likelihood of this happening by chance would hinge on the extent of the bandwidth—i.e. on the inherent characteristics of the weapon—and on the nature of the ecosystem. For example, an inverse correlation is to be expected with the species-diversity of the ecosystem. The likelihood of it happening by design will no doubt increase as knowledge in systems-ecology advances. Chemical and biological weapons, in particular, offer a disturbingly accessible species-selectivity in their biocidal actions.

The importance of a species of organism within an ecosystem may be assessed by reference to function. For example, the more a species influences the energy throughput of an ecosystem, or the more strongly it affects the environment of all other species, the more important will it be. In these respects, particular species within particular ecosystems may be identified as "ecological dominants". On the criterion of energy throughput, the ecological dominants within any one trophic group will be those that have the greatest productivity.

For any ecosystem, the most obviously vital category is that of the primary producer organisms: the green plants which, through the mechanism of photosynthesis, are the ecosystem's principal means of converting solar energy into the chemical energy necessary for sustaining life and growth in its other biotic components. Thus, the stress of biocide upon an ecosystem may be especially strong if a significant range of plant species falls within the biocidal bandwidth of the weapon concerned. On pages 50 and 51 we have indicated the magnitude of some of the possible consequences.

The consequences may assume their gravest proportions in the case of forest destruction. So powerful is the momentum which forests may impart to the cycling of nutrients, and so strong may their influence be on the hydrological and meteorological conditions of a region, that their importance within an ecosystem, and hence their loss, may be felt over an area far exceeding that which is actually forested. There have been innumerable instances in history, which continue to repeat themselves even now, of deforestation being succeeded by widespread land degradation and impoverishment of the region, amounting, in some cases, to desertification and other forms of ecological disaster.

Ecological dominants tend to become fewer in number as geographical latitude increases. In a northern forest, for example, 90 percent or more of the total stand may be made up of one or two species of tree, whereas in a tropical forest it may comprise a dozen or more species. In this particular respect, tropical forests may thus be vulnerable to a stress of narrow biocidal bandwidth than a northern forest.

As to the consequences of biocidal stress upon species at higher trophic levels, it is not yet possible to identify with precision depopulations that would inevitably have an adverse impact upon ecosystems as a whole. Within several ecosystems more narrowly defined that those of the present study, dependencies of one species upon another, as in the foodchains or the pollinating insects, can, of course, be specified; and there are many reported observations of the elimination of one population in an area provoking strong oscillations in other populations.

RELATIONSHIPS BETWEEN SCALES OF WEAPONS-USE AND MAGNITUDES OF STRESS AND STRAIN

The strain which an ecosystem may experience when subjected to any of the foregoing stresses will depend upon the characteristics of the ecosystem. It will also depend upon the magnitude of each stress, which in turn will depend upon the nature of the weapons concerned and the scale on which they are used. For each possible weapon-ecosystem impingement, the strain may be conceived in terms of a threshold below which the ecosystem vulnerability, in the sense mentioned on page 32, is insignificant. For some types of weapon and ecosystem, this threshold may be dangerously low. For others, the threshold may be so high as to be unsurpassable during military operations of a plausible kind and scale.

The scale on which weapons may be used will depend upon a range of circumstantial factors of which the limiting one is, ultimately, the availability of resources to the belligerents. A more immediate limiting factor is the number of committed delivery systems available and suitable for the weapons in question. A very rough idea of possible magnitudes of each type of stress that might realistically be anticipated may therefore be gained by estimating for one type of common weapon-delivery-system the area over which the different stresses might be created were a full complement of each of the different weapons to be delivered by it. The resultant area-estimates may then be converted into minimum numbers of weapon-delivery-systems that could stress an area of a given size. Some such estimates, for an area of 500 km^2, are set out in Table 4. The delivery-system chosen is a ground-attack aircraft having a payload of up to 6 tonnes of ordance, its munitions delivery capacity being configured so that it could carry two munitions of 1000-kg rating. Such aircraft are to be found in the air forces of many countries, though in widely differing numbers.

On page 47 above, a theoretical approach has been suggested which could be used to relate the rough measures of stress suggested by Table 4 to the thresholds of significant strain envisaged above. The approach requires that damage caused by stress due to weapons acting upon the structural components of an ecosystem be assessed in terms of the extent to which the functional processes that couple the structural components are thereby affected. The degree to which overall function is impaired then appears as the measure of strain. If the strain is manifest as harm to human populations in the sense mentioned on page 32, the threshold of significant strain will have been crossed. The ecosystem will have proved brittle, not resilient, in the face of the weapon insult.

It has to be said, however, that the full capacity of such an approach for differentiating the likely ecosystemic impacts of the various types of weapon hinges upon knowledge that does not yet appear to exist. Much is known about particular aspects of both structure and function in a variety of

TABLE 4. ECOSYSTEM-STRESSING POTENTIALS OF DIFFERENT WEAPONS
Rough estimates of the numbers of aircraft[a] carrying each type of weapon that might expose 500 square kilometres of terrain to different types of ecosystemic stress

Weapon Class	Type	Stress factor				Biocide	
		Soil damage		Scorching[b]	Plant-cover destruction[c]	Producer organisms[d]	Consumer organisms[e]
		Displacement					
High-explosive	General-purpose HE	340 000[k]		—	7 200[l]	?	7 200[l]
	Fuel-air explosive[f]	—		?	2 300	?	?
Incendiary	Napalm	—		14 000	<42 000[m]	?	?
Chemical	Nerve gas[g]	—		—	—	—	1 100[n]
	Herbicide[h]	—		—	1 700[o]	600[p]	—
Biological	Plague bacteria	—		—	—	—	>30
	Wheat-rust fungi[i]	—		—	—	>3[q]	—
Nuclear	20 kt standard fission[j]	27 000		10[r]	70[s]	9 - 300[t]	9[u]
	1 Mt fission-fusion-fission[j]	2 000		2[r]	3[s]	1 - 3[t]	1[u]

[a] Ground-attack aircraft each capable of delivering up to 6 tons of munitions. It is further assumed that, for the high-explosive and incendiary munitions, the payload would comprise bombs of 200–350 kg rating; that the chemical and biological agents would be delivered from spraytanks comparable in size to bombs of 1000 kg rating, two per aircraft; and that no more than two nuclear munitions would be delivered by each aircraft.
[b] Refers to soil insufficiently vegetated for wildfire propagation.
[c] Refers to forest or other woody vegetation.
[d] Refers to death of crop-plant species.
[e] Refers to 50 percent mortality among populations of man or similar species.
[f] Refers to hypothetical third-generation fuel-air explosive munitions.
[g] Refers to o-ethyl S-2-disopropylaminoethyl methylphosphonothiolate.
[h] Refers to a formulation comprising equal parts by volume of the n-butyl esters of 2,4-dichloro- and 2,4,5-trichlorophenoxyacetic acids.
[i] Refers to uredospores of *Puccinia graminis tritici*.
[j] Fused for surface-burst.
[k] Delay-fused bombs.
[l] Quick-fused bombs.
[m] Lower figure if a spreading wildfire is initiated.
[n] A lethal hazard may persist in the area for several weeks or more.
[o] Refers to at least 60 percent defoliation lasting for 4-14 weeks of multi-layered deciduous-forest canopy with peak defoliation within 10-24 weeks.
[p] Refers to broad-leaved food-crop cultivations.
[q] Refers to 85 percent loss of wheat yield if weapons are used under conditions most favouring epiphytetic spread.
[r] Refers to thermal irradiation.
[s] Refers to 90 percent tree-blowdown due to blast.
[t] Refers to nuclear radiation, the range reflecting radiosensitivities of different crop-plant species.
[u] Refers to 30-day mortality from nuclear radiation.

ecosystems. But the ignorance remaining is such that any attempt to *correlate* structure and function must deal in hypotheses which will rarely, on the data at present available, be either substantiable or refutable. For the concluding sections that now follow, in which are presented overall impact appraisals guided by the foregoing analytical approach, this is a major caveat.

IMPACTS ON TROPICAL ECOSYSTEMS

It was noted on page 32 that a major characteristic of tropical ecosystems, most apparent in the tropical rainforest, is that their cycling nutrients tend in comparison with those of other ecosystems to have longer residence times in the biomass than in the geomass. This is presumably due to a greater rapidity in the heterotrophic utilization of the products of autotrophic organisms. The more the cyclic commodities of an ecosystem are concentrated in the biomass, the more vulnerable may the ecosystem be to biocidal stress. The lesser role of the geomass in nutrient cycling is likely to mean that it will prove incapable, by reason of adaptations among the biotic community, of filling breaches in the cycle caused by biocide. The quality of the soil, for example, is likely to be such that nutrients entering it via the decomposers from a sudden overload of dead plant or animal organisms will not be retained for long, being more or less rapidly leached away by rainfall. There may then occur the phenomenon of "nutrient dumping", already alluded to, in which the entire ecosystem suffers a continuing loss of nutrients. Productivity would fall at an increasing rate unless or until recovery via recolonization started to recouple the broken circuits of the nutrient cycle. In the worst case the ecosystem could suffer desertification.

A tropical ecosystem will be most endangered via nutrient dumping if biocidal stress is concentrated in regions of especially high biomass. Thus weapons may have a particularly strong impact on tropical ecosystems if they are used within heavily vegetated regions and if their biocidal bandwidth encompasses a high proportion of the species comprising that vegetation. The rapidity of the biocide, and its phasing in relation to seasonal changes, including inflexions in rainfall, will be major contributory factors. For example, if the biocide is relatively slow, the increase in the nutrient load on the geomass may be spread over a sufficiently long time for it to be adequately retained. In such cases nutrient dumping may not occur significantly unless either or both of the other two major stress factors operate. Plant-cover destruction, in the sense of soil denudation, would increase exposure of the soil to the leaching and erosive actions of heavy rainfall or winds; so would soil displacement.

Nuclear weapons most obviously pose the threat of nutrient dumping to tropical ecosystems. High-explosive weapons might do so, at least in theory, but it seems clear from Table 4 that very large numbers of weapon-delivery systems would have to be employed for such effects to be anything other than highly localized. Incendiary weapons would seem to present either a lesser threat or an intermediate one. The latter would be the case if the fires which they initiated took hold and spread of their own accord. As noted on page 18, there would be a substantial probability of this happening in regions where natural wildfires are common. Here, however, the ecological dominants would presumably be adapted to survive wildfires, even to prosper in their aftermath. This would not be so in wildfire-rare regions, which would be correspondingly more vulnerable, but here the incendiary attack could have to be of peculiar intensity to initiate a wildfire. Yet given the advances that are now being made in the technology of incendiary weapons, particularly flame weapons, this is a possibility that cannot be discounted.

Herbicidal chemical weapons would also present an intermediate threat. If both biocide and soil-denudation resulted from their use over a large enough area, catastrophic nutrient dumping might well occur. But it would seem from Table 4 that such effects would require a large commitment of delivery systems for the requisite chemical weapons. As noted on page 53, the ecological dominants of tropical vegetation—at least of the kind of vegetation against which antiplant chemical weapons might, on military logic, be employed—are relatively large in number. The species diversity would necessitate a wide biocidal bandwidth from the herbicides employed if the soil was to be significantly denuded. If high-explosive or incendiary weapons, especially the latter in the less humid regions, were used in conjunction with herbicides, the danger of nutrient dumping might substantially increase.

The stress factors of soil damage, plant-cover destruction and biocide could operate to menace tropical ecosystems by mechanisms other than nutrient dumping. The sharply regionalized danger of soil laterization and consequent desertification has been noted on page 48. Some solace may be derived from the fact that laterization has not been reported from areas of Indochina in which latosols are present and which were subjected to military herbicide operations and other forms of weapons insult. But since it has occurred in certain parts of tropical South America as a consequence of forest clearance, it is a hazard that cannot be ignored.

Other serious consequences of weapons-impact on tropical forest can be envisaged. The risk of soil induration follows from one of a number of fragilities in tropical soil. As noted on page 53, forests may shape structure and function in ecosystems, sustaining productivity, over areas distant from those that are actually forested. Such areas may include ones given over to agriculture. In the relatively poor soils common in the tropics, agriculture may be precarious at the best of times, but if adjacent forest vegetation is destroyed, agricultural productivity may seriously decline. Increased erosion, which is always a possibility after deforestation, may act upon lands already virtually denuded by the demands of agriculture and, in especially vulnerable areas, desertification may follow.

Under conditions of high ambient temperature, moisture increases in importance as a limiting factor for most organisms. As the amount of water falls below the limits of tolerance for forests, trees begin to give way to grassland and the like. The labile equilibrium that exists between woody and grassy plant species in tropical savannah has been noted on page 37. Weapons that expose tropical savannah ecosystems to stresses to which the grassy vegetation is especially sensitive may have the effect of reducing transpiration, thereby tilting the equilibrium so that there occurs an invasion of woody species, initially in the form of scrub or thorn. The value of the land for grazing purposes will then be reduced. If the woody vegetation is harvested for firewood or thorn-fencing, the result will be an aggravated denudation of the land, probably followed by erosion and hence desertification. Weapons possessing the discriminatory biocidal action necessary to induce such a train of events would seem to be limited to chemical and biological weapons having a species selectivity for grasses. All other types of weapon would be likely to affect either both sides of the equilibrium alike, or to act more strongly on the woody species.

IMPACTS ON ARID ECOSYSTEMS

As the climate becomes progressively drier in hot regions, the limiting factor of moisture becomes displayed in a diminishing species diversity. Food-chains become shorter, and less variety is available for consumer organisms in their choice of food. This imposes an increasing lability on the equilibrium between the different populations that constitute the biotic community of arid ecosystems. Exposure to the biocidal stress of weapons, especially ones of relatively narrow biocidal bandwidth, may therefore throw the equilibrium into more or less violent oscillation. The indigenous human population of the ecosystem may not have the resources for protection against the consequences and in the worst situation may itself go into rapid decline. The aftermath of the Sahelian drought, in which the grazing lands could no longer support sufficient livestock, is a terrible illustration of such a process.

There are reasonable grounds for doubting whether any use of non-nuclear weapons could induce population oscillations of a kind having human consequences comparable to that of the Sahelian disaster unless that use were on a scale so massive as to serve no other purpose. Only the use of nuclear weapons might result in such an effect unintendedly, and even then the numbers of weapons used might have to be large enough to encompass entire nomadic areas within their overall zone of herbicidal action. In such a case the direct damage to the human population might dwarf the ecosystemically mediated damage.

Though it seems improbably, it could be that real dangers of such a magnitude, albeit spread over a longer time period, exist in the more significant role which small consumer organisms appear to play in arid ecosystems as compared with others (see page 36). Taken together with the lability attendant upon low species diversity, this factor, if indeed it exists, could magnify the overall ecosystemic impact of biocidal weapons specific for such creatures. Such a possibility could presumably come about only with weapons capable of generating fine aerosols of nerve gas or of particular toxins or microbial pathogens. But, unless used at night and on a scale that ensured a sufficiently wide area coverage by the aerosol before the daytime lapse conditions in atmospheric stability set in, such an aerosol would most likely fail to permeate the small-consumer habitat to an adequate extent. If there is any empirical evidence on this matter from arid-region weapon-test sites, it has yet to be published.

In arid regions moisture is both an upper and a lower limiting condition upon most plant species. But although an increase in moisture may, because of the abundance of incident solar energy, lead to a rapid colonization by species adapted to greater moisture, a decrease in moisture does not lead to an equally rapid colonization by species less demanding of moisture. Thus, as noted on page 37, soil damage which reduces moisture retention may precipitate a sudden loss of plant cover, which in turn will expose the soil to further degradation by wind erosion. This is a mechanism of desertification which, in some soil types, might be set in motion by heavy use of high-explosive or flame weapons. Since the eroding topsoil could be carried in dust storms, vegetation outside the affected area might also be destroyed.

It is conceivable that such a desertification mechanism might be activated the other way about, with partial plant-cover destruction causing a decline in the moisture-retention of the soil sufficiently great to ensure complete soil denudation, and sufficently long lasting to deny regrowth of the same moisture-adapted plant species; erosive wind action might then permanently impoverish the soil. Any of the weapons capable of stressing ecosystems by plant-cover destruction or herbicidal biospecificity might set such events in motion. Among non-nuclear weapons, the relatively great area effectiveness of anti-plant chemical weapons would be the most threatening, though it would be hard to envisage a military rationale other than ecocide for their use in arid ecosystems.

IMPACTS ON ARCTIC ECOSYSTEMS

Arctic ecosystems have vulnerabilities towards weapons-stress in common with arid ecosystems because of their poor species diversity. Much of what is said in the preceding section may be no less applicable to them.

They have additional vulnerabilities that stem from low ambient temperature. Primary productivity is low and the biogeochemical cycling of nutrients is very slow. Biocidal stresses of weapons acting upon the primary producer communities of arctic ecosystems as a whole and a very long-lasting one. Recovery periods may stretch over several human generations, so that major losses of vegetation may, for practical purposes, amount to desertification. Tundra forest, once lost, may never recover.

The cold may act to increase greatly the environmental persistence of toxic agents disseminated by chemical weapons in arctic ecosystems. Their biocidal action may thus become considerably extended by virtue of the prolonged time periods over which they place sensitive organisms at risk. The sluggishness of the biogeochemical cycling will reduce the mobility in the ecosystem of any toxic or radiotoxic substances that enter the food-chains. However, since the latter are abnormally short, with scant variety of foodstuffs for consumer organisms, there may nonetheless be vicious manifestations of bioaccumulation in the case of some poisons and radiounclides. Thus, although the lability of the population equilibria defining arctic communities might well be thrown into oscillation by any large-scale use of weapons in arctic ecosystems, it can be expected that the fluctuations of population will be markedly more violent and long-lasting in the cases of nuclear weapons and some types of chemical weapon.

Chapter 5

CONCLUSIONS AND RECOMMENDATIONS

OVERALL IMPLICATIONS

It is useful to restate briefly certain broad themes. Man as an individual depends for survival, fundamentally, upon the shelter and nourishment derivable from his environment. For human populations, the same is true, but at this level human well-being also requires the existence of favourable balances with other populations of living organism: animal and plant species that provide food; species upon which those foodstuff populations themselves depend, directly and indirectly; pathogenic micro-organisms and species predatory upon man or his livestock; and so on. It is in these balances, and in man's growing capacity for influencing them to his advantage, that the prospects for equitable social and economic development are rooted. And it is on those balances that the security of the different subgroupings of the global human population ultimately depend, including the national security of individual nation-states.[49] So long as the nature of the balances remains roughly predictable, providing tolerable assurance of continuity, there will be a rational basis for pursuing the paramount social goals of development and international security. Conversely, upheavals in the balances may jeopardise all such aspirations and set regions of the world into a conflict-ridden future, ravaged, in the worst case, by wars which few nations may be able to escape.

To the extent that war may itself cause such upheavals, the balances could

---

[49]For a cogent expression of this theme, see Lester R. Brown, "Redefining national security", *Worldwatch Paper series* (Washington, D.C.: Worldwatch Institute) No. 14, October 1977.

CONCLUSIONS AND RECOMMENDATIONS

be poised for an accelerating deterioration of the human prospect.[50] This is the direst of the possibilities in considering the effects of weapons from the ecological standpoint. An ecosystemic approach of the type used in this study would be an apt means of appraisal, compelling attention, as it does, to dynamic equilibria among populations and between them and their physical environment. The objective of the present study has been more limited, its underlying concern being with desertification. This, however, is a phenomenon which could prove, in those arid, tropical and arctic regions to which most consideration has been given here, a heavy contributing factor, perhaps the heaviest, to collapse of the overall balances. It is estimated that between 50 and 78 million people are directly affected by decreases in productivity associated with current desertification processes.[51]

DESERTIFICATION

Several possible mechanisms have been identified whereby ecosystemic impacts of weapons might culminate in the desertification of particular regions. The danger of this happening appears by far the greatest with nuclear weapons. In the most vulnerable regions, the danger may also arise with several other classes of weapon, including ones not customarily classified as "weapons of mass destruction". The factors which may impart such vulnerability to a region include, in particular, lack of species diversity and thinness, or poor nutrient- or water-holding capacity, of soil. One or the other, or both, of these commonly occur in tropical, arid and arctic regions.

Nuclear weapons present the gravest risk of desertification both because of the greatness of the area over which their effects may be manifest and because the nature of their effects may trigger any one of several of the different desertification mechanisms identified in this study. The mechanisms include: nutrient-dumping, which is a possibility in tropical ecosystems of high primary productivity following heavy mortality in the biomass combined with soil-denudation; soil induration, which might be induced by destruction of plant cover in those tropical regions where lateritic soils occur; and accelerating soil erosion which, in arid or sparse regions, might be induced by weapons in

---

[50] For an analysis of the possible roles of war and war-preparedness in conceivable futures, see M. H. Kaldor and J. P. Perry Robinson, "War", in C. Freeman and M. Jahoda (eds.), *World Futures: The Great Debate* (London: Robertson, 1978), pp. 343-370.
[51] A/CONF.74/1/Rev.1 at para. 31.

several ways, such as by mechanical or thermal damage to the soil that reduces water retention or, indirectly, by species-selective mortality that overloads foodchains. Most of these are mechanisms which weapons having a strong biospecific action may be especially likely to set in motion, particularly weapons having a biospecificity for the plant species that dominate primary production in a region and therefore also soil cover. This is a circumstance which, as regards desertification, justifies particular concern about some forms of chemical and biological weapon.

The mechanisms of possible desertification outlined above, and specified in more detail earlier, fall into two broad categories. In the one, desertification would be confined to the area immediately affected by the weapons. In the other, the damage in that area might not itself constitute desertification but its perturbation of ecosystem function might instead cause a slow desertifying process gradually to build up within the general region. The mechanism would be one of positive feedback loops taking hold within the damaged ecosystem, possibly increasing in number as the desertifying process advances, maybe activating such wider factors as albedo change. The insidiousness of desertification feeding upon itself in this fashion is particularly to be feared.

In the more vulnerable regions, forests may play an especially important role within the larger ecosystems of which they form parts. Their importance for the nutrient cycling, hydrology and meteorology of the region, and its stability in the face of natural erosive forces, may be such that deforestation could, in the worst case, precipitate desertification. In less vulnerable regions, the ecosystemic effects of deforestation might still be strongly felt, for example as reduced overall productivity in adjacent unforested areas. It is important to appreciate, also, that the management of ecosystems by man, as in agriculture, may have the effect of increasing their vulnerability, for land given over to crop-plants grown in monoculture is, in ecological terms, very nearly bare. Agriculture in forested regions necessitates some deforestation, and, when the demands on agriculture are large, the situation is usually reached in which deforestation has been pushed to the limits that agriculture can tolerate. Any further deforestation may therefore have a sharply adverse impact on the human population. Agricultural communities in tropical areas, and wider economies dependent upon them, could be especially endangered. Deforestation may be brought about by the action of most types of weapon used in large enough quantities. For chemical weapons of the herbicidal type, and for incendiary weapons in the drier forest, especially the newer types of

CONCLUSIONS AND RECOMMENDATIONS

flame weapon, the quantities may be substantially smaller than for other weapons, excepting nuclear weapons.

Agriculture is only one form of management by man of the natural environment. In fact there are rather few types of human endeavour that cannot be regarded as man seeking to manage ecosystems of which his own population is a consumer component. At this level of abstraction, however, the meaning of desertification loses clarity. For example, urbanization may in one sense be seen as a form of desertification. In another sense it may be seen as an adaptive habit on the part of the human species that contributes to the overall function of the ecosystem populated by urban man, in which case it would be the mass-destruction, not the growth, of urban areas that might become analogous to desertification. In a third sense—one much closer to the normal—weapons-assault on human artefacts such as dams or large-scale desalination plant could trigger desertification. The denser the human population of a region, the more intensely may it suffer from the consequences of damage to managed ecosystems, whether or not they are mediated by desertification.

MAGNITUDE OF IMPACT

Desertification of the self-propagating kind distinguished on page 64 is one of several forms of damage which, through ecosystemic function, may become transmitted to contiguous regions. Other such kinds of spreading damage may be identified among the conceivable impacts of weapons upon ecosystems. The broader consequences of deforestation other than desertification equally noted on page 64 are one illustration. Another is the damage that might be caused by massive volumes of particulate material carried by wind or water away from areas of erosion: damage in the form of, for example, silting or accelerated eutrophication of aquatic ecosystems. Climate change, including increased ultraviolet irradiation, such as might follow large-scale use of nuclear weapons (see page 26) are a third.

Such possibilities would complicate attempts to measure, let alone predict, the overall magnitude of the ecosystemic effects of the employment of weapons in particular types and quantities in particular regions. Ability to make such assessments, at least in semiquantitative terms, seems to be presupposed in the current provisions of international law applicable in armed conflict that relate to environmental protection. These provisions have been noted earlier, on pages 3 and 4. The expression "widespread, long-term and severe damage" is used in Article 35 (3) of 1977 Geneva Protocol I; and "widespread, long-lasting or severe effects" is used in the 1977 Environmental Modification Convention. It may be recalled that during the negotiation of the Environmental Modification Convention, the following agreement was unanimously reached at the Conference of the Committee on Disarmament:

> "It is the understanding of the Committee that, for the purposes of this Convention, the terms 'widespread', 'long-lasting' and 'severe' shall be interpreted as follows: (a) 'widespread': encompassing an area on the scale of several hundred square kilometres; (b) 'long-lasting': lasting for a period of months or approximately a season; (c) 'severe': involving serious or significant disruption or harm to human life, natural and economic resources or other assets. It is further understood that the interpretation set forth above is intended exclusively for this Convention and is not intended to prejudice the interpretation of the same or similar terms if used in connexion with any other international agreement."[52]

Though one of the objectives of the present study has been to explore the scientific and technical bases whereupon such interpretations might rest, it has not been possible to reach firm conclusions. The stress-strain model suggested on pages 42, 53 and 54 may perhaps have some merit, if indeed quantification serves a useful purpose. However, while it may be possible to quantify, along the lines suggested on page 54 and illustrated in Table 4, the degree of stress that a given expenditure of weapons may place on ecosystems, it will remain extremely difficult to quantify the resultant strain in the ecosystem and to specify the manner and scale on which that strain might become manifest as damage to the natural environment or as "destruction, damage or injury" to a State. What would be needed for this is further descriptive information about the ecological profiles of the more obviously vulnerable regions.

There is in any case a strong requirement for the latter type of research, for the information it provides would also serve as the basis for early warning networks in the protection of ecosystems which non-military forms of human activity may already be placing in jeopardy.

FURTHER STUDY

The conclusions which may be drawn from the present study are limited by

---

[52] Conference of the Committee on Disarmament, Document CCD/520, Annex A.

CONCLUSIONS AND RECOMMENDATIONS 67

its nature, which is that of a preliminary mapping-out exercise. This limitation affects not only the firmness of the conclusions reached but also their scope. In particular, the criterion of ecosystem vulnerability (page 32) on which the impact assessments rest, though reasonable enough for an exploratory survey, is too crude to capture the full range of adverse impacts that weapons may have on ecosystems. Strictly applied, it confines attention to ecosystems delimited broadly enough to subsume man within their biotic components. An important reason why this is more constraining than would otherwise be desirable is that most of current ecological knowledge and understanding of ecosystem function relates to smaller ecosystems. Moreover, this vulnerability criterion excludes from consideration indirect impacts of weapons via those dependencies characterizing international and intersocietal relations in which ecosystems become linked through patterns of exploitation, import and export. Weapons impacting upon an ecosystem at one end of such a relationship could have a most adverse effect upon the human population of the ecosystem at the other end. To take an extreme example, the criterion denies vulnerability of oceanic ecosystems; yet mass-destruction weapons, through their possible effects on the structure or function of such ecosystems, could have enormously bad consequences for human populations that are directly or indirectly dependent upon marine resources. The consideration given to temperate ecosystems likewise needs to be extended.

These limitations are noted for the bearing they may have on any follow-up study that is undertaken. In this regard, three other matters should be taken into consideration. The first is that the methodological weaknesses on the present study are not limited to those noted on page 9; above all, for an investigation cutting across several different scientific disciplines and areas of knowledge, it is important that each should be adequately represented. In parts of the present study—places, for example, where there is more structure than content—the need for additional intellectual input is apparent. The second matter, which relates somewhat to the choice of vulnerability criterion discussed on the previous page, is the need to set an assessment of the possible ecosystemic impacts of weapons within a perspective broad enough to accomodate other adverse impacts of man's activities on the environment. There may be important synergies that would otherwise escape attention: synergies in overall impact between, for example, war and environmental pollution or depletion of nonrenewable resources.

The third matter is that of the security needs of States as currently perceived by them. Such needs find expression in the maintenance and equipment of armed forces, including the types of weapon that are acquired. A pertinent parallel exists in the manner in which security demands have influenced the development of international humanitarian law applicable in armed conflict. This influence has been a strongly restraining one, and it must be expected that similar restraints will affect analogous attempts to develop international environmental law, especially if the global imperative of environmental protection is insufficiently appreciated. It may be necessary to recognize that, just as there are gradations in military necessity, so also are there gradations in the ecosystemic impacts of weapons: one way forward would then to be to seek ways in which the two might be correlated.

RECOMMENDATIONS

In summary, two broad conclusions may be drawn from this preliminary study. The first is that, while modern weapons are capable of causing enormous environmental damage anywhere in the world, some parts of the global ecosystem appear to be considerably more brittle than others. In such regions, this lack of resilience in the web of interdependencies and interactions which link human societies to one another and to their natural environment may greatly magnify the damaging effects of weapons. The magnitude and precise character of this risk cannot be specified without more detailed study. But it seems clear that the risk exists; that it may introduce dangerous assymmetries into a world where conflicts continue; and that, because the component ecosystems of the biosphere are all closely interdependent, it represents a danger to the entire world against which there is scant protection or remedy.

The second broad conclusion follows from the first. If additional safeguards are to be developed against the danger, there must first be greater awareness within the international community of its existence, and its character must be more fully investigated. It is most strongly to be recommended that action be taken to implement both objectives. There is now extensive concern about the environment in public opinion which may strongly favour initiatives on safeguards. But whether the initiatives are taken within the disarmament negotiating machinery, as most properly they might, or in any other international forum, the necessary preparatory work must be undertaken and completed before the present timeliness passes.

ANNEX

FIELD INVESTIGATIONS INTO THE ECOLOGICAL SEQUELAE OF MILITARY HERBICIDE OPERATIONS IN INDOCHINA: A BIBLIOGRAPHY

Sponsoring Organisation	*Published Reports*
US State Department (1968)	F. H. Tschirley, "Defoliation in Vietnam", *Science* **163**: 779-786 (1969).
US Society for Social Responsibility in Science (1969)	G. H. Orians and E. W. Pfeiffer, "Ecological effects of the war in Vietnam", *Science* **168**: 544-554 (1970); and E. W. Pfeiffer and G. H. Orians, "Military uses of herbicides in Vietnam" in J. B. Neilands *et al.* (eds.), *Harvest of Death: Chemical Warfare in Vietnam and Cambodia* (New York: Free Press, 1972), pp. 117-176.
US Scientists' Committee on Chemical and Biological Warfare (1969)	A. H. Westing, "Herbicidal damage to Cambodia", ibid; pp. 177-205.
American Association for the Advancement of Science (1970)	J. Constable and M. S. Meselson, "Ecological impact of large scale defoliation in Vietnam", *Sierra Club Bulletin* **56** (4): 4-9 (1971); M. S. Meselson *et al.*,

Sponsoring Organisation	Published Reports
	Preliminary report of the Herbicide Assessment Commission of the American Association for the Advancement of Science", in *War-related Civilian Problems in Indochina. I. Vietnam* (Committee on the Judiciary, US Senate, 1971), pp. 113-131; A. H. Westing, "Ecological effects of military defoliation on the forests of South Vietnam", *BioScience* **21**: 893-898 (1971); and A. H. Westing, "AAAS Herbicide Assessment Commission", *Science* **179**: 1278-1279 (1973). The final report of the AAAS/HAC is still in preparation.
US Scientists' Institute for Public Information (1971)	E. W. Pfeiffer, *Ecocide: A Strategy of War* (a 21-minute 16 mm film by Thorne Films of Boulder, Colorado, 1972); and A. H. Westing, "Herbicides in war: current status and future doubt", *Biological Conservation* **4**: 322-327 (172).
US Defense Department (1971-1973)	A Lang et al., *Effects of Herbicides in South Vietnam* (US National Academy of Sciences, 1974, 20 vols).
UN Special Mission (1977)	C. O. Lankester, part IV ("Forestry and forest industries: war damage and development prospects") of the Report of the Special Mission on International Assistance for the Reconstruction of Viet Nam Appointed by the Secretary-General in Accordance with General Assembly Resolution 32/3 (UN document Control No. 78-02689).